ALSO BY MARK V. STEIN

Machine Age to Jet Age: Radiomania's Guide to Tabletop Radios 1933-1959 Vol. I

Machine Age To Jet Age
Volume II

Machine Age To Jet Age Volume II

Radiomania's® Guide to Tabletop Radios
1930-1959
(with market values)

Mark V. Stein

In Collaboration with
Alan Jesperson and Mike Emery

Radiomania® Publishing
2109 Carterdale Road
Baltimore, Maryland 21209

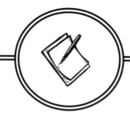

NOTICE

Published by
Radiomania® Publishing
Copyright 1997

Additional copies of this book may be purchased or ordered
from your local specialty bookseller, or directly from the publisher
by sending $28.95 (or $24.95 for Volume I) by check or money order payable to:

Radiomania® Publishing
Department 2
2109 Carterdale Road 21209 USA

(US shipping free, Canada add $2.00, all other countries add $3.00)

E-mail: Radioman@cais.com   Fax: (410) 466-0815
Website: http://www.sscsi.com/machine_age/radiomania

Future editions of this reference are currently in progress.
If you care to contribute photographs, advertising brochures or other items,
please write the author directly in care of the above address.
All contributed items will be returned upon request and contributors given
due credit in the acknowledgments

Book designed by Jane E. Rubini

Dedication

This one's for my folks,
Leonard and Rona Stein.
For all the times they dragged me along
with them to those musty old
antique barns.

Most importantly,
for their continual confidence and
support in whatever direction I chose.
Thanks Mom and Dad.

ACKNOWLEDGEMENTS:

THIS BOOK WAS REALIZED THROUGH
THE COOPERATIVE EFFORTS OF MANY
COLLECTORS. THE MAJOR CONTRIBUTORS
ARE AS FOLLOWS:

ALAN JESPERSON AND MIKE EMERY
RADIO COLLECTORS AND PROPRIETORS OF 'GREAT NORTHERN',
A VINTAGE RADIO RETAIL AND MAIL ORDER STORE SPECIALIZING IN
ZENITH AND E.H. SCOTT FLOOR MODEL RADIOS, REPRODUCTION
KNOBS, SUPPLIES, ADVERTISING AND MEMORABILIA
GREAT NORTHERN
PO BOX 17338, MINNEAPOLIS, MN 55417
(612) 727-2489

JOE GREENBAUM
RECOGNIZED NATIONAL AUTHORITY ON VINTAGE RADIO CABINET
RESTORATION AND REFINISHING. PROPRIETOR OF GREENBAUM
REFINISHING, AND AUTHOR OF NUMEROUS ARTICLES ON THE SUBJECT
GREENBAUM REFINISHING
312 S. EXETER ST.
BALTIMORE, MD 21202
(410) 752-2438
HTTP://WWW.EROLS.COM/GMNINS/GREENBAUM/GRNBM1.HTM

JAY KIESLING
COLLECTOR SPECIALIZING IN RADIOS BY ATWATER KENT
AND UNUSUAL CATHEDRAL STYLE RADIOS
15906 TRENTON ROAD, UPPERCO, MD 21155
(410) 239-1818

DONALD EDDY
COLLECTOR SPECIALIZING IN UNUSUAL 1930S RADIOS INCLUDING
GLOBE DIALS, HIGH TUBE COUNT, MOTORIZED TUNING AND IN
1950S VARIABLE SELECTIVITY RADIOS
2521 W. NEEDMORE HWY., CHARLOTTE, MI 48813
(517) 543-3021

IN ADDITION TO THE ABOVE, THE FOLLOWING INDIVIDUALS HAVE ALSO
CONTRIBUTED TO THE COMPLETION OF THIS BOOK:

IRA GROSSMAN, RICHARD BOSCH, MIKE STAMBAUGH, FRANK MOORE,
ALAN VOORHEES, DAVID MEDNICK, DOUG HEIMSTEAD, JIM CLARK,
JIM MEEHAN, RICHARD LOBAN, TED DEPTO, RADIO BILL FROM DETROIT,
JOHN SAKAS, STEVEN SANDLER, BOB KAMINSKY, JAY MALKIN, BEN MARTIN

THANKS TO ALL!

Machine Age To Jet Age
Volume II

Radiomania's® Guide to Tabletop Radios 1930-1959

TABLE OF CONTENTS

Machine Age To Jet Age
Volume II

Radiomania's® Guide to Tabletop Radios 1930-1959

TABLE OF CONTENTS

Machine Age To Jet Age
Volume II

Radiomania's® Guide to Tabletop Radios 1930-1959

TABLE OF CONTENTS

Machine Age To Jet Age
Volume II
Radiomania's® Guide to Tabletop Radios 1930-1959

TABLE OF CONTENTS

PREFACE

Welcome to the second volume of Radiomania's vintage radio reference guide. Due to the tremendous response to Volume One of this series, I am please to present to you a second and supplemental Volume Two more than two years in the making. In compiling this second volume reference, I have tried to stay true to the intent of the first in focusing on a single, high profile area in vintage radio collecting, the tabletop radios of the high style era beginning in the early 1930s and continuing through the 1950s. I have slightly broadened the time frame in Volume Two to include the years 1930-1932, dominated by what we now call cathedral or beehive type tabletop radios. As previously mentioned, this reference serves as a supplement to Volume One, and is comprised of all new listings so that the radios pictured in both volumes combined total over 4,000. This is, by far, the most comprehensive reference book on the subject and I am far from finished. That's right, Volume Three is already in the works.

In compiling this text, I continued to use collectors and collector meets as resources for subject photography. In addition, through the generosity of advanced collectors such as Alan Jesperson, Mike Emery, Jay Kiesling, and Donald Eddy I have been able to tap large resources of original advertising and collateral industry materials to further broaden the scope. I continue to search for additional and related information valuable to the collector and thank Joe Greenbaum for his time and efforts in authoring the chapter related to wood cabinetry. My hearty thanks goes out to these and all hobbyists involved in the development of this book. I encourage others to share their resources with fellow collectors through the continuation of this series. Your thoughts, comments and contributions are welcomed.

Thank you and enjoy!

Mark V. Stein

PRICING

In compiling this book the question as to whether to include prices at all was one of major concern to both the author and contributors. As most collectors are aware, even the best price guide, if not initially flawed, is soon obsolete. The range of prices paid for most items is wide. One can pay anywhere from a few dollars to a few hundred for a given item, dependent on where it was purchased and from whom. Thus the task of establishing a 'market value' is difficult if not impossible. This problem given, it was still generally felt that the value of including prices, at least as a general market gauge, would outweigh the disadvantages of omitting them completely.

In establishing values a number of sources were utilized including auction results, classified ads, meet pricing, collector valuations and the author's personal experience having been both a collector and dealer for over ten years. Prices in this book represent items in average condition. Average meaning that the radio is intact but 'as found'. No repairs would have been made, either electrically or otherwise. If wood, the finish would be original. It might show some wear, but would be presentable in a collection once cleaned up. If plastic, the cabinet would be free from cracks and chips. Plaskon radios in average condition might evidence some minor stress lines but nothing which would detract from their aesthetics. Values followed by a '+' symbol represent rarer items, only a few of which have traded hands. In such cases the value listed represents an estimate of what a collector might expect to pay for an average example. The '+' symbol is also used to indicate baseline values for catalin radios, the price of which will vary widely based on color, condition, marbelization, and variations. This reference does not attempt to address the intricacies of valuing catalin radios, but merely to provide the reader with a baseline gauge.

A word about dealers and dealer prices: expect to pay a premium when purchasing from a dealer. The dealer offers one the luxury of eliminating the time consuming hunt through yard and estate sales, flea markets, antique shows and the like. It is he who goes through the trouble of rooting out those hard to find items. Ones which you might not happen upon except after years of hunting yourself. Dealers inventories represent long hours and related expenses and must reflect those additional costs.

FINE TUNING VALUES

To assist the collector in fine tuning the value of a given radio, some general rules of thumb are offered. Please remember that, as with all rules, there are exceptions to these. If you are unsure about a specific item you will always do best to ask another collector whom you trust.

GENERAL CONSIDERATIONS

CHASSIS CORROSION: In buying any radio it is important to visually inspect the chassis, particularly if you plan to restore the set to working order. Minor surface corrosion is typical and acceptable to all but the most finicky of collectors and should not deter from the value. Extreme corrosion can be an indication that the radio was submerged at one time or at least has seen a lot of humidity. If so, the bulk of the internal components may need to be replaced. Even if you plan only to display the radio, if it is so damaged its value will be decreased by the fact that it would be less desirable to most other collectors.

RODENT WEAR: Most of us have, at some time, come across a radio with leaves stuffed into the nooks and crannies of its chassis and acorns which have mysteriously gotten underneath, or a set which has had every wire, both exposed and internal, chewed through. Such phenomena are caused by common vermin such as squirrels, rats and mice. At one time the radio was probably stored outside in a barn or garage and it happened to become someones home. The damage caused by such occupancy can be severe. If you come across such a set and are tempted to buy it, inspect it carefully. The easily observed superficial problems may be but a sign of more extreme damage internally. Yank the chassis if you plan to restore it to working order. You may chose not to after you see the innards.

TRANSFORMERS: A common problem, and one that may be costly to remedy, is that of the 'smoked' transformer. Caused by a short circuit or overload, damage to the transformer almost always necessitates its replacement. Fortunately, such problems are often easy to spot. Look for smoke damage to the chassis and on the inside of the cabinet. Also look for gooey stuff (which may have hardened over time) oozing from the transformer itself. Most collectors stay far away from sets with blown transformers if at all possible. Short of finding a junker radio of the same variety but with a good transformer, you'll need to buy a replacement with the same specs.   This can cost anywhere from twenty to ninety dollars assuming you are lucky enough to even find a match.

TUBES: Missing tubes are generally easy to replace, particularly amongst radios from this era. The resources are numerous. Your local club will probably be your least expensive resource (usually one to three dollars per tube). After that there are local repair shops, mail order companies and specialty houses. It is a good idea to know your resources when considering the purchase of a radio sans tubes. Also be aware that there are critters known as ballast tubes which look like filament tubes superficially but which are really resistors. Count on about one out of every two ballast tubes needing replacement. Unlike filament tubes, few ballasts have survived. Replacing a bad ballast will require much perseverance or some clever electronic rigging to bypass the device.

BACKS: Most radios pictured in this book had backs when they were originally sold. Most don't now. The backs were typically flimsy cardboard which became worn and fell off never to be seen again. Often times the antenna was attached to the missing back and is now also missing. To most collectors the absence of a cardboard back will not detract from the value of a radio. Conversely, the presence of a back, particularly if it is in good condition, will add value. One exception to this rule are those radios made during the mid to late 1930s with moulded plastic backs (like the Fada 260 series or the Emerson 199). With these radios the back is considered to be a critical element of the radio design and, as such, its absence can devalue a set by as much as 25%.

KNOBS: If one or more knobs are missing from a radio its value will be diminished. The purchaser of such a radio must resolve himself to either completing the set or substituting another complete set that looks right. If you chose to complete the set with correct knobs, the most cost effective, but also the most time consuming, method is to bring a correct knob with you to a local collector meet in hopes of finding a match in a box of odd knobs for sale. Cost for such knobs is typically in the one to five dollar range. Be prepared to wait a long time and look through a lot of boxes though. Another way to complete the set is to purchase a reproduction from one of several vendors throughout the country. Check ads in trade periodicals and club newsletters for such resources. There are dozens of types being reproduced and stocked today. The price for such reproduction knobs will range from two to seven dollars. Specialty knobs with chrome rings or those that duplicate catalin can run upwards of twenty-five dollars each. If you've got an odd bird which is not being reproduced, your final alternative is a specialty vendor who will make a casting of your knob with reproductions upon order. Cost for such knobs typically begins at ten dollars. Regardless of how you decide to deal with the missing knob(s), the value of the radio will be reduced by an amount which correlates directly with the time, cost and
aggravation of completing the set.

DIAL LENSES: Covering the dial area on most radios pictured in this book is a plastic or glass dial lense. A damaged or missing plastic dial lense can be reproduced to order by a number of vendors. The going rate is about fifteen dollars per lense. If you don't have the original you can still make a tracing of the dial opening and send it to with your order. Again, look in periodicals and club bulletins for resources.

Glass dial lenses are typically made from convex round pieces of glass secured either inside the dial bezel or attached to the dial scale on the radio itself. Clockmakers usually will have matches for such pieces in the form of clock crystals. Just measure the outside diameter of the glass and make a few calls to area clockmakers or contact a supply house. You might, as an alternative, decide to replace the glass with plastic and contact a lense reproducer.

Both glass and plastic dial lenses were sometimes reverse painted with dial scales and ornamentation. If such a dial is broken, cracked, warped or otherwise damaged you've got a significant problem facing you. Except for a limited number of high value radios, no one makes reproduction dial scales. It will be next to impossible to find a suitable replacement. Your best hope will be to find another of the same radio and use its glass. You'd probbly do just as well to wait for that next radio instead. The only reasons to buy a radio with such a problem are either that the price is extremely reasonable (enough to make the problem tolerable) or that you do not expect to have another opportunity to own such a radio in the forseeable future.

PLASTICS

For the novice colllector the number of different plastics used in the manufacture of radio cabinets can be confusing. Furthermore, the names we use today have been established more from convention than anything else. That is, most all of the descriptive names currently used were at one time a brand or product name for the material. Typically, many manufacturers produced the same or a similar material. For one reason or another a single brand name has, over time, become synonymous with the material itself. The most common types of plastics and their conventions are as follows:

BAKELITE: The 'first' plastic, Bakelite was made from a powder and molded under tremendous pressure and heat. It is typically brown or black and somewhat pourous. Brown bakelite can be heavily 'mottled' with various shades of coloration.

BEETLE: Formed, like bakelite, under heat and pressure, this early plastic can be striking in appearance. Used to mold cabinets from the late 1930s through the early 1940s, beetle is typically opaque ivory in color with marbled streaks of orange, rust, geen, blue, red and brown. Some examples may be subtle with just a hint of rust marbelizing while others evidence themselves in a wide array of deep color tones.

CATALIN: An early resin, catalin was poured into molds and then cured in low heat ovens over a period of days. Once removed from the mold, if not broken in the process, the catalin radio cabinet was machined to remove debris and to add detail to the design. The labor intensity of this material was not at issue in the late 1930s when it was first introduced in radio cabinetry, but after World War II, with the shortage of manpower that resulted, its use quickly became cost prohibitive.

Catalin is known to be the most valuable of plastics used in radios with colorations ranging from opaque solid colors to semi-transluscent marbles resembling polished agate. Catalin is more fragile than other plastics of the period and is prone to chips, cracks, burns (from tube heat), discoloration (those butterscotch radios were originally white) and shrinking. All of these hazards have had a significant impact on the number of catalin sets which have survived, thus driving up their price.

PLASKON: Also manufactured using a process similar to bakelite, plaskon was molded in ivory and opaque colors. Typically, radio manufacturers offered the consumer the option of an ivory color at a higher price. Sometimes this was simply a painted brown or black bakelite cabinet. Other times it was molded in ivory plaskon. Occaisionally, radios were offered in colors other than ivory. Plaskon cabinets were available in colors such as pistaschio green, chinese red and lavender. Although plaskon wears about as well as bakelite, it is subject to stress lines. These are superficial cracks which, unfortunately, fill with dirt and grime over the years. Some can be cleaned or bleached out, others can be sanded down, but most must be lived with.

PLASTIC: Refers to contemporary injection molded plastics, most frequently polystyrene. First widely used in the late 1940s and available in a variety of colors.

TENITE: Occaisionally used for cabinets, tenite was most often used for the manufacture of radio grilles, knobs, handles and ornamental parts. The most significant problem with this material was its vulnerability to warpage due to proximity to heat. Tenite was not a particularly wise choice for use in construction of heated filament tube-type radios.   Almost all tenite warps to one degree or another. The best one can hope for is minimal warpage.

VALUING PLASTIC RADIOS

CHIPS AND CRACKS: As a rule of thunb, a major flaw in any plastic radio, such as a significant visible chip, crack or warp, will cut the value of that radio in half so long as it remains displayable. Among plaskon radios hairline stress cracks are fairly common. In many models they are the rule and not the exception. As stated previously, so long as the stress lines are not too numerous and do not detract materially from the general aesthetics of the radio, the depreciation should be minimal. On the other hand, the stress free example of a set which is commonly found with stress lines is worth considerably more than average.

FRAGILITY: Dependent on the thickness of the casting, materials used in construction and the extremity of design, some radios are inherently more delicate than others. The most fragile (such as the Kadette 'Classic') are rarely found in near perfect condition. This 'universe' from which each radio is drawn must be considered in determining its value. In our example, a Kadette 'Classic' in what would appear to the casual observer as marginal condition, might, in fact, be an excellent example of that model, given the condition of other surviving sets.

REPAIRS: Although there have been some relatively successful attempts at bakelite repair, no repair can go undetected. A well repaired flaw can increase the value of a radio but will never raise its value to that of an unflawed one.

PAINT CHIPS: While many of these early plastic radios were available in different colors (typically walnut, black and ivory), oftentimes the ivory colored sets were brown or black bakelite which had been painted at the factory. This factory painting process was similar to that used in the automobile industry. After several layers of paint were applied, the cabinet was baked for several hours. As a result, the paint became 'hard' and much more susceptible to chipping over time. In addition to the problem of chipping, the baking process made the paint that much more difficult to remove. Stripping old baked on paint is extremely difficult and time intensive at best. Be prepared to use caustic chemicals and spend a lot of time. You may then find yourself with a black or brown radio and cast ivory plaskon knobs. This may look fine to you, but it is not 'original' and will be less valuable to other collectors. The net result is that the value of a factory painted radio with paint chips will be reduced by anywhere from ten to fifty percent dependent on the extent and placement of chipping.

VALUING WOOD RADIOS

Although typically not as popular with collectors in the past, wood cabinet radios are gaining in value and popularity as the hobby expands. It is interesting to note that, when radios were sold in the 1930s and 1940s, it was the wood cabinet which sold for a premium. Plastics were considered then to be a cheap man-made alternative.

The following are some basics to keep in mind when considering the purchase of a wood cabinet radio:

FINISH: Most wood radios produced during the 1930s through the 1950s were finished with either clear or toned lacquer. The more expensive sets were hand rubbed resulting in the 'piano' finish many today find so desireable. Over the years many things can and usually do happen to a lacquer finish. The finish will dry and chip, peel or separate ('alligator'). Rarely does one come across an original lacquer finish without some evidence of the passage of time. Unless the surviving finish is horrendous, it is worthwhile trying to salvage. There are various products available which will allow you to easily clean the surface and replenish the moisture in the lacquer. Additionally, you may want to amalgamize the finish to cover bare spots. This involves dissolving the original finish with denatured alcohol or other substance and respreading it.

As opposed to clear lacquer, toned lacquers present more of a problem. Many manufacturers, instead of using different wood veneers for contrast, used a single veneer type and sprayed toned lacquers for variations in color and shade. Originally, the difference was difficult to discern. Today it is evident. When toned lacquers age they reveal the original color of the wood underneath, usually in deep contrast. The darker the toner, the more significant the problem. The purchaser of such a radio must resign himself to live with the radio as is or refinish it. Both are a compromise.

VENEERS: Most all early radio cabinets were made with wood veneers as opposed to solid exotic or hardwoods. Over time the glue which bonds the veneer to the wood of the cabinet can deteriorate. The result ranges from small veneer chips to full pieces of veneer falling off. The best case is where the veneer is separating, but still present. This is easy to remedy with glue and clamps. If pieces of veneer are actually missing, they must obviously be replaced. This means removing the remainder of the damaged section of veneer and replacing it with another full piece. You may be able to find such a piece with an original finish on a junker radio. Otherwise, you will need to apply a finish to the new veneer. If the piece is very small and in a relatively concealed area you may chose to fill the hole with a matching wood puttly, shellac stick or toner pencil.

Additionally, veneer tends to lose its flexibility over time. Veneer which has been bent at angles of ninety degrees or more on a radio may break at those turns. If this occurs, again, be prepared to live with it or replace that entire section of veneer.

PAPER VENEER: A concept similar to the toned lacquer, paper veneer involved application of exotic veneer decals to the radio cabinet. This technique was used for both whole radio cabinets and small details. Upon close inspection of the veneer, one can usually discriminate between paper and real veneers. A damaged decal will make the difference apparent. If a radio is ornamented with damaged paper finish, any attempt to repair or refinish the radio will likely remove the decal or damage it further. This should be kept in mind when evaluating such a set for purchase.

REFINISHING: All but the absolute purists will agree that some radios just need to be refinished in order to be displayable. Many collectors have not seen a well refinished radio and so cannot appreciate the art to it. The process is long and involved, and extremely time intensive, but the result of a 'professional' refinish is astonishing. Of course there is refinishing and there is refinishing. Many collectors will use a caustic solvent to remove the original finish, then stain the cabinet and apply a coat of polyurethane or tung oil. The result is something that looks like it came out of a craft shop, not a vintage radio. If you are considering the purchase of a radio which is so refinished it should be valued as if it had no finish at all (about half the value of such a set in average condition).

Other types of refinishing are less objectionable and easier to remedy. Oftentimes one encounters a radio with the original finish intact. It is just underneath a coat of slopped on varnish or shellac. Various solvents can be used to remove the top coat and leave the original finish. There is an art to this process in both identifying the right solvent and its application. Read up on refinishing and talk to other collectors about their experiences before attempting this on a valuable acquisition. The net result on the value of radios so 'refinished' is a reduction by about twenty-five percent. A note of caution: be certain that the slopped on coat is not polyurethane. If it is, you're back to square one and must remove all finishes.

WOOD RADIO FINISHES

By Joe Greenbaum

FINISHES: A finish serves two primary purposes: first, to enhance the beauty of the wood, and; second, to protect it. Contrary to casual perception, a traditional lacquer finish involves much work and preparation and typically is comprised of several steps including staining, grain filling, sealing and application of many light top coats of lacquer.

Often times, during the first half of the twentieth century, in the mass production of wood items such as radios, some of the traditional steps in the application of a lacquer finish were skipped or replaced with more 'efficient' alternatives. For example, in order to save time, labor and materials expenses, toned lacquers were often used. Toned lacquers combined both the stain and top coat processes into one by adding staining pigment to the first few top coats of lacquer. This allowed manufacturers to use inferior quality and/or unmatched woods which, if traditionally stained, would not be presentable. The toned lacquer replaced the stain and would lie on top of the wood instead of penetrating it. This had the effect of masking the imperfections and variations and would blend them to look uniform. To the layperson toned lacquers could make an inexpensive poplar cabinet look like walnut, mahogany or other high cost hardwoods.

In a tradional lacquer finish the process of grain filling is key. Grain filling involves filling the pits created by pores in the raw wood and accentuated by staining with a semi-liquid putty, letting it harden and then sanding it smooth. Again, in the mass production of radios, this process was often ommited, particularly when less expensive woods with less pronounced grain and, therefore, smaller pores, were used. Instead, heavier or more numerous top coats were applied which, when completed, would leave a relatively smooth and presentable product.

Once the top coats were applied, the more expensive lines of radio cabinetry were hand-rubbed with progressively finer abrasives to a 'piano' finish. The majority of radio cabinets, however, were not hand rubbed. Instead, commercial spray equipment was able to lay a relatively smooth top coat which, for retail purposes, was saleable without further polishing.

Like any finish, time takes its toll on lacquer. Even the finest lacquer finishes may craze, crack, peel, flake, chip or discolor over time. Factors which may effect the survival of a finish include proximity to high temperatures such as caused by radio tube heat, extreme variations in temperature and/or humidity such as caused by storage in a barn or attic, and, of course, contact with water and other solvents such as caused by setting drinks or canisters on top of the surface. When evaluating a vintage radio for restoration/refinishing it is always best to retain the original finish, or at least part of it, if at all possible. There are various products available which will clean, almalgamate (dissolve and redistribute), and restore aged lacquer finishes. Make no mistake, a correctly refinished wood cabinet can be worth as much as a well preserved original, however, there is something to be said for historical preservation, not to mention the time, effort and expense of replacing the original finish.

A FEW WORDS ABOUT REFINISHING: I'd like to start by reiterating that radios were originally finished with lacquer top coats, not with varnish, shellac, polyurethane, tung or linseed oil. Refinishing with non-original materials and/or methods will significantly decrease the value of the radio to other collectors, even if the end result is acceptable to you. If you are going to refinish at all, it is proper to use the materials and methods originally used manufacturing the cabinet. If you are not sure how to properly apply a lacquer finish it is best to leave it alone or pass it on to someone who does. An improperly applied lacquer finish not only looks bad but will also significantly decrease the value of the radio. There are several reference books available to the novice who would like to learn lacquer finishing. Additionally, many local wood supply stores and some refinishers will offer hands-on courses. If you plan to refinish yourself make sure that you have a significant amount of time to spend on the process. As mentioned previously, proper refinishing is extremely time consuming, even to a professional. This is the primary reason it is so expensive when you have a professional provide the service. That having been said, the end result can be breathtaking. The depth of fine hardwood and exotic veneers is unparalled by any manmade material. A well finished radio cabinet can be priceless to its owner.

If you have decided to maintain an original finish, it is best to first make certain that the finish present is, in fact, original. You can do some detective work using a few basic solvents: lacquer thinner; denatured alcohol; and naphtha. First find an area of the radio which is not easily visible, like a back corner, on which to test and clean it with some naphtha applied to a cotton cloth. Then pour some denatured alcohol on another cotton cloth and rub it repeatedly on the test area. If the finish dissolves it most likely shellac. Shellac was often applied by radio shops directly over the lacquer finish to 'rejuvinate' it. If this is the case, you can use denatured alcohol on the entire radio and rub the shellac finish completely off. This will oftentimes leave you with an intact and well-preserved original lacquer finish since the denatured alcohol will not dissolve the lacquer. If, however, the denatured alcohol does not dissolve the finish then try some lacquer thinner. If it dissolves the finish you can bet that its lacquer. Remember to always try denatured alcohol first because lacquer thinner will dissolve both shellac and lacquer finishes. Finally, if neither solvent dissolves the finish you've got either varnish or polyurethane, in which case you have no choice but to strip. There is no way to preserve the original finish.

If you do not have an original finish and have decided to strip, proceed cautiosly. Try non-caustic chemicals for stripping first. They may be slower working and require several applications but caustic stripping chemicals such as methyl chloride can dissolve the glues holding the veneers down and the cabinet together creating a much worse situation, not to mention the health hazards of such chemicals. Once the finish is removed you may begin the application of the new finish in ernest and that is a subject worthy of a book in itelf.

Joe Greenbaum is the sole proprietor of Greenbaum Refinishing located in Baltimore, Maryland. He is nationally renowned for his fine quality historical restoration of vintage radio and clock cabinets in addition to his general furniture refinishing business. Joe has spoken at national radio meets and published in several periodicals. He also conducts refinishing courses at his shop. He may be contacted at 312 S. Exeter St., Baltimore, MD 21202, (410) 752-2438. Website: http://www..erols.com/gmnins/greenbaum/grnbm1.htm

RESOURCES

As the hobby of vintage radio collecting has evolved over time, so have the resources for collectors. There are regional, national and international clubs. Cottage industries have sprung up offering a wide range of products and services. There are club newsletters, bulletins, references books and a monthly periodical on the subject. The following pages serve to provide the novice with some basic resources. This is, by far, not a comprehensive listing of all resources available. Any ommisions are not purposeful and do not reflect in any way on the individuals or organizations omitted.

VINTAGE RADIO CLUBS:

By far the best part of the hobby is meeting and spending time with other collectors. Club meets provide a forum for exchange of information, purchase of material and new radio acquisitions among other things. There are few collectors in the United States who are not within a few hours drive from a local or regional club. The following is a listing of regional, national and international clubs. When writing any organization for information be sure to include a postage paid return envelope to assure a response.

REGIONAL CLUBS (UNITED STATES)

Alabama Historical Radio Society
PO Box 26452, Birmingham, AL 35226

Arkansas Chapter/AWA
PO Box 191117, Little Rock, AR 72219

Antique Radio Club of Illinois
RR3, 200 Langham, Morton, IL 61550

Arizona Antique Radio Club
2025 E. La Jolla Dr., Tempe, AZ 85282

Antique Radio Club of Ft. Smith
7917 Hermitage Dr., Ft. Smith, AR 72903

Belleville Area Antique Radio Club
219 W. Spring, Marissa, IL 62257

Antique Radio of Greater St. Louis
103 Pleasant St., Jerseyville, IL 62052

Buckeye Antique Radio Club
4572 Mark Trail, Copley, OH 44321

Antique Radio Collectors of Ohio
PO Box 292292, Kettering, OH 45429

California Historical Radio Society (CHRS)
PO Box 31659, San Francisco, CA 94131

CHRS-North Valley Chapter
15853 Ontario Pl., Redding, CA 96001

Hudson Valley Antique Radio/AWA
PO Box 207, Campbell Hall, NY 10916

Carolina Chapter/AWA
1236 Autumn Oaks Dr., Lancaster, SC 29720

Indiana Historical Radio Society
245 N. Oakland Ave., Indianapolis, IN 46201

Central PA Radio Collectors Club
1440 Lafayette Parkway, Williamsport, PA 17701

Iowa Antique Radio Club & Historical Society
2191 Graham Cir., Dubuque, IA 52002

Cincinnati Antique Radio Collectors
5916 Ropes Dr., Cincinnati, OH 45244

Louisana & Mississippi Gulf Coast Area
640-18 Tete LKours, Mandeville, AL 70471

Colorado Radio Collectors
5270 E. Nassau Circle, Englewood, CO 80110

Memphis Antique Radio Club
7190 Mimosa, Germantown, TN 38138

Connecticut Area Antique Radio Collectors
500 Tobacco St., Lebanon, CT 06249

Michigan Antique Radio Club
2590 W. Needmore Hwy, Charlotte, MI 48813

Connecticut Vintage Radio Club
70 Litchfield Rd., Unionville, CT 06085

Mid-America Antique Radio Club
80006 Greenwald, Belton, MO 64012

Delaware Valley Historic Radio Club
PO Box 41031, Philadelphia, PA 19127

Mid-Atlantic Antique Radio Club
PO Box 67, Upperco, MD 21155

East Carolina Antique Radio Club
218 Bent Creek Dr., Greenville, NC 27834

Mid-South Antique Radio Collectors
2811 Highland Ave, Carrollton, KY 41008

Florida Antique Wireless Group
Box 738, Chuluota, FL 32766

Midwest Radio Club
PO Box 6291, Lincoln, NE 68506

Four States Antique Radio Club
1702 S. McKinley Ave., Joplin, MO 64801

Mississippi Historical Radio & Broadcasting
2412 C St., Meridian, MS 39301

Greater Boston Antique Radio Collectors
12 Shawmut Ave., Cochituate, MA 01778

Mountains 'N' Plains Radio Collectors
1249 Solstice ln., Fort Collins, CO 80525

Greater NY Vintage Wireless Association
52 Uranus Rd., Rocky Pt., NY 11778

Music City Vintage Radio & Phono Society
PO Box 22291, Nashville, TN 37202

Houston Vintage Radio Association
PO Box 31276, Houston, TX 77231

Nebraska Antique Radio Collectors Club
905 West First, North Platte, NE 69101

New England Antique Radio Club
PO Box 918, Melrose, MA 02176

Sacramento Historical Radio Society
PO Box 162162, Sacramento, CA 95816

New Jersey Antique Radio Club
2265 Emerald Park Dr, Forked River, NJ 08731

Society of Wireless Pioneers
146 Coleen St., Livermore, CA 94550

New Mexico Radio Collectors Club
1371B Baker Dr. SE, Albuquerque, NM 87116

Southeastern Antique Radio Society
PO Box 500025, Atlanta, GA 31150

Niagara Frontier Wireless Association
135 Autumnwood, Cheektowaga, NY 14227

Southern California Antique Radio Society
6934 Orion Ave., VanNuys, CA 91406

Northland Antique Radio Club
Box 18362, Minneapolis, MN 55418

Southern Vintage Wireless Association
3049 Box Canyon Rd., Huntsville, AL 35803

Northwest Vintage Radio Society
PO Box 82379, Portland, OR 97282

South Florida Antique Radio Collectors
1717 N Bayshore Dr #1251, Miami, FL 33132

North Jersey/New York/Long Island
Box 172, Valley Cottage, NY 10989

SPARK/Cincinnati Chapter
PO Box 81, Newport, KY 41071

North Chapter/CHRS
15853 Ontario Pl, Redding, CA 96001

SPARK/Columbus Chapter
2327 E Livingston Ave, Columbus, OH 43209

Oklahoma Vintage Radio Club
PO Box 332, Wheatland, OK 73097

Texas Antique Radio Club
7111 Misty Brook, San Antonio, TX 78250

Pittsburgh Antique Radio Society, Inc.
83 Ruthfred Dr, Pittsburgh, PA 15241

Tidewater Antique Radio Association
2328 Springfield, Ave., Norfolk, VA 23523

Puget Sound Antique Radio Association
PO Box 125, Snohomish, WA 98291

Treasure Coast Antique Wireless Club
1445 2nd Rd SW, Vero Beach, FL 32962

RI Antique Radio Enthusiasts
61 Columbus Ave., N. Providence, RI 02911

Triangle Antique Radio Society
8490 Hurdle Mills Rd, Hurdle Mills, NC 27541

Radio Enthusiasts of Puget Sound
14911 Linden North, Seattle, WA 98133

Vintage Audio Listeners
1127 NW Bright Star Ln., Poulsbo, WA 98370

Radio Historical Society
4147 Lenox Dr, Fairfax, VA 22032

Vintage Radio & Phonograph Society
PO Box 165345, Irving, TX 75016

Vintage Radio Unique Society
312 Auburndale St., Winston-Salem, NC 27104

Western WI Antique Radio Collectors
1611 Redfield St., La Crosse, WI 54601

West Virginia Chapter/AWA
405 8th Ave., St. Albans, WV 25177

Xtal Set Sociaty
789 N. 1500 Rd., Lawrence, KS 66049

U.S. NATIONAL AND INTERNATIONAL CLUBS:

Antique Wireless Association (AWA)*
Box E, Breesport, NY 14816

International Antique Radio Club
PO Box 5261, Old Bridge, NJ 08857

FOREIGN CLUBS:

AUSTRALIA

Historical Radio Society of Australia
PO Box 283, Mt. Waverly, Victoria 3149

CANADA

Canadian Vintage Radio Society/CVRS:
British Columbia Chapter
4895 Mahood Dr., Richmond, BC V7E 5C3
Manitoba Chapter
3216 Assiniboine Ave, Winnipeg R3K 0B1

London Vintage Radio Club
19 Honeysuckle Cres., London N5Y 4P3

Ottowa Vintage Radio Club
Box 84084 Pinecrest PO, Ottowa K2C 3Z2

Quebec Society for Vintage Radio
799 S. Etienne St., Granby, Quebec J2G 9N8

ENGLAND

British Vintage Wireless Society
23 Rosendale Rd.
West Dulwich, London SE21 8DS

FRANCE

Club Histoire et Collection Radio
26 Rue de l'Oratoire, 54000 Nancy

French Antique Radio Association
135 Av. du President Wilson, 93100 Montreuil

GERMANY

German Society of Wireless History
Belm Tannenhof 55, 7900 Ulm 10

HOLLAND

N.V.H.R.
19, 6814 K.T., Ahmen

IRELAND

Irish Vintage Radio & Sound Society
39A Lower Drumconda Rd., Dublin 9

ISRAEL

Antique Radio & Broadcast Museum
24 Remez St., #7, Tel Aviv 62192

ITALY
Associazione Italiana Radio d'Epoca
Via de Pellicceria 23, 52100 Arrezo

NEW ZEALAND

New Zealand Vintage Radio Society
20 Rimu Road, Mangere Bridge, Auckland

NORWAY

N.R.H.F.
PO Box 465 Sentrum, N-0105 Oslo 1

SPAIN

Freiends of Radio Cultural Assn.
c/o Rei Jaume, 55, 08840 Cardedeu

SWEDEN

Radio Historical Society
Gata 2, 417 55 Guteborg

The preceeding listing was provided courtesy of <u>Antique Radio Classified</u>
See Publications section for further information about this periodical.

Periodicals:

Antique Radio Classified is published monthly and is the foremost periodical dedicated to the hobby. Issues typically run in excess of 100 pages and include articles by collectors addressing technical issues, collecting, historic perspectives, test equipment, etc. Other regular features include a photo review of unusual items, meet and auction reports, club meeting notices, events calendars and, of course, a huge classified section. Sample articles and general collecting and club meet information can be found at the ARC web site (www.antiqueradio.com) or free sample copies can be obtained by writing Antique Radio Classified, P.O. Box 2-V79, Carlisle, MA 01741.

Radio Club Periodicals are published monthly, quarterly or less often by most radio clubs for their membership. The largest such periodical is published by the AWA (Antique Wireless Association) on a quarterly basis. Additional information available directly from each organization. See club listings.

Reference Guides:

Other books are available which provide a number of perspectives on the radio collecting hobby. These include the following:

Classic Plastic Radios of the 1930s and 1940s, Sideli
E.P. Dutton, 2 Park Avenue, NY, NY 10016

Evolution of the Radio, Volumes 1 and 2
LW Books, PO Box 69, Gas City Indiana

A Flick of the Switch 1930-1950, McMahon
McMahon Vintage Radio, Box 1331, North Highlands, CA 95660

Golden Age of Radio in the Home and More Golden Age of Radio, Stokes
Craigs Printing co., Ltd., 67 Tay Street, Invercargill, New Zealand

Philco Radio 1928-1942, Ramirez & Prosise
Schiffer Publishing Ltd., 77 Lower Valley Road, Atglen, PA 19310

Radio Art, Hawes
Green Wood Publishing Co., Ltd., 6/7 Warren Mews, London W1P 5DJ

Zenith Radio Brochure Book, Jesperson & Emery
Great Northern, PO Box 17338, Minneapolis, MN 55417

With the recent advent of the Worldwide Web, shopping and resourcing for the hobbyist is now much more convenient. Although the internet remains in its infancy, with sites coming and going constantly, there are a few 'supersites' worth mentioning even now. Listed below are some such representative websites:

ANTIQUE RADIO CLASSIFIED: http://www.antiqueradio.com/toc.html
Sponsored by the largest vintage radio periodical includes club & event listing, article reprints, large list of radio related website links

ANTIQUE RADIO COLLECTOR: http://www.members.gnn.com/richmann1/wrldradio/index.htm
Free classifieds, radio related articles, listening room

ANTIQUE RADIO GRILLECLOTH: http://www.libertynet.org:80/~grlcloth/
Sponsored by John Okolowicz, includes visual samples of all styles of replacement grillecloth plus great information on radios attributed to industrial designers.

ANTIQUE WIRELESS ASSN.:   http://www.ggw.org/freenet/a/awa
Website for the largest national club in North America. Includes primarily information about the AWA museum in Bloomfield, NY.

GOLDEN AGE OF RADIO: http://alvo.kqed.org
*Information on old radio shows and related subjects, on-line museum, radios for sale.
Sponsored by San Francisco radio station KQED.*

GREENBAUM REFINISHING: http://www..erols.com/gmnins/greenbaum/grnbm1.htm
Sponsored by nationally renouned radio refinisher. Includes photos of recent projects and opportunity to ask questions via e-mail with previous Qs & As posted.

MACHINE AGE: http://www.sscsi.com/machine_age
20th Century Design Superstore/Mall - Includes several on-line stores, three of which (Radiomania, Radio Craze, New Era Antiques) specialize in the sale of vntage radios and related items. Also included are free on-line classifieds, Exhibition Information, Museums, links to other related websites.

MARC :   http://www.sojourn.com/~micharc/web/index.html
Excellent site sponsored by Michigan Antique Radio Club (MARC), host of one of the largest national radio events, Extravaganza. Site includes on-line newsletter, classifieds, meet reports and great links.

ON-LINE RADIO TRADER: http://www.http://footnet.com/collect/main.htm
Free computer searchable classifieds for antique radios and related items.

RADIOMANIA®:   http://www.sscsi.com/machine_age/radiomania
By far the largest on-line vintage radio store, selling vintage radios, books, and related industrial design artifacts. Sponsored by Mark V. Stein, the author of the Machine Age to Jet Age references.

RCA STUDIOS: http://www.mcs.net/~richsam/home.html
Sponsored by Chicago's Merchandise Mart, great historic look at NBC Radio Studios in Chicago with tours and historic photos/accounts of the premises and links to similar sites for historic television.

As mentioned previously, the above is a mere sample of the great websites out there. Most all sites have links to other related sites. Once you start exploring you'll be amazed at the quantity and variety of this new a growing resource.

There are a host of suppliers and vendors who will make your life as a collector a little easier. Listed below are a representative few:

Antique Electronic Supply, 6221 S, Maple Avenue, Tempe, AZ 85283
Wholesale supplier of vintage tubes, electronic components, reproduction knobs, grille cloth and other items.

Constantine's, 2050 Eastchester Rd., Bonx, NY 10461 (718) 792-1600
Complete line of woodworking tools and supplies including veneers.

Great Northern, PO Box 17338, Minneapolis, MN 55417
Wide range of NOS and reproduction Zenith parts and other items. Write for catalog.

Greenbaum Radio Refinishing, 312 S. Exeter St., Baltimore, MD 21202
Complete range of professional wood cabinet refinishing. (410) 752-2438.
http://www..erols.com/gmnins/greenbaum/grnbm1.htm

Michael Katz, 3987 Daleview Ave., Seaford, NY 11783
Reproduction radio grille cloth. Send $.52 LSASE for samples.

Kotton KleanserProducts, Inc., PO Box 1386, Braden, TN 38010-1386
Manufacturer of wood cleaner, protective wood feeder and other antique resoration products.

Vintage Radio & TV Supply, 3498 West 105th Street, Cleveland, OH 44111
Reproduction knobs, parts, tubes and other items. (216) 671-6712

Wades World of Knobs, 7109 E. Arbor Ave., Mesa, AZ 85208
Catalog of reproduction knobs and castings made to order. Write for catalog and information.

Doyle Roberts,   HC63, Box 236-1, Clinton, AR 72031
Dial lense reproduction from damaged old lense or drawing of dial opening.

COLOUR PLATES

At the request of many readers of the first volume in this series, I have undergone the initiative (and expense) of including colour plates in this volume. The following pages represent selected examples of various types of tabletop radios. Classification is by cabinet material and/or cabinet style, one usually being a function of the other. Pictured radios are further described and valued in either this book or Volume 1 of the 'Machine Age to Jet Age' series.

BAKELITE

FADA 260

EMERSON 21A

CROSLEY B548A

SETCHELL-CARLSON 485R

DEWALD

MANTOLA 461-5SL

STEWART-WARNER 07-5B

GE H-500

FIRESTONE R-320

MOTOROLA 79XM21

SPARTON (CAN) PEE WEE

STROMBERG-CARLSON 1500H

BAKELITE: PAINTED

CROSLEY D-25BE

PHILCO 49-501

STEWART-WARNER 9182

SKY CHIEF 'PATRIOT'

CROSLEY E-10

SILVERTONE 3061

CROSLEY 11-115U

CROSLEY JT3

GAROD 5A2

CROSLEY 11-101U

CROSLEY 56TD

STEWART-WARNER 9162

Beetle Plastic

Detrola 281

Addsion 2

Silvertone 3361

GE H510

Kadette 'H'

Sentinel 195ULTO

Silvertone 3351

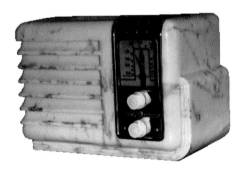

GE GD520

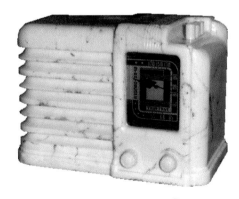

Truetone 278-5Q

Detrola 283

Sonora TW49

Admiral 4202-B6

PLASTIC

CONTINENTAL 1600

BENDIX 114

RCA 4X551

TELE-TONE TR70

PHILIPS (CAN) B1C12U

CROSLEY 11-113U

STANDARD

EMERSON 744B

WESTINGHOUSE H583-T5

GENERAL TELEVISION

FADA 1005

PHILCO J775-124

Urea

Admiral Y-300

Emerson 517

Motorola 57CS

Motorola 56H

Epsey 174

Emerson 707B

Motorola 57X

Westinghouse H679-T4

Emerson 826B

Motorola 56C5

Emerson 724

Motorola 57H

PLASKON I

Air King 871

Kadette 'Classic'

Emerson 108

Emerson 157

Detrola 219

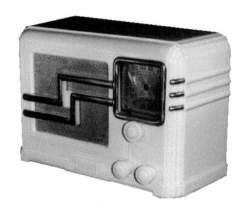

Fada 454

RCA 96X

Setchell-Carlson 416

Emerson 149

Fada

Airline 05GCB-1541

Westinghouse WR-166

PLASKON II

LIONEL TRAINS PROMO

DETROLA 280

AIR KING A-600

WILCOX-GAY A-53

RCA 9SX

RCA 9TX

EMERSON 211

DETROLA 218

DEWALD

PHILCO

SILVERTONE 'CORONET'

MAJESTIC 'ENSIGN'

CATHEDRALS I

JACKSON BELL 68

SIMPLEX R

LARK

JACKSON BELL 84

CORONADO

CROYDON

ATWATER KENT 944

PHILCO 70

STEINITE 16

US RADIO 32

STEWART-WARNER R110

PHILCO 20 DELUXE

CATHEDRALS II

STEIN 'AZTEC'

DeFOREST CROSLEY (CANADA)

ZANEY-GILL 'VITA-TONE'

ACRATONE 132

OZARKA 93

DeWALD 740

COLONIAL R6764

STUDEBAKER 'WREN'

ATWATER KENT 735

MAJESTIC 196

ATWATER KENT 558

JACKSON BELL 32

METAL GRILLES I

JEWEL R188

STEWART WARNER 'COMPANION'

WILCOX-GAY A-15

AIR KING

DEFOREST RADIO INSTITUTE

MAJESTIC 149

MAJESTIC 49

EMERSON 31A

ARKAY

MAJESTIC 461

BEVERLY (MACY'S)

ZENITH 809

Grunow 450

Belmont/Freshman 426

Stewart Warner 1235

Majestic 59

Lyric C3/M4

Grunow 500

Zenith 829

Colonial 301

Majestic 174

Majestic Extension Speaker

Pla-Pal 'Penthouse'

Zemith 835

Novelties

Globetrotter 'Globe'

Pure Oil Promtion

Trophy 'Baseball'

General Television 'Piano'

Radio-Glo

Sparton 557 'Sled'

Abbotwares Z477

RCA 40X-53 'La Siesta'

Wings Cigarettes Promotion

Philco 53-706

RCA 'Bookset'

General Tire Promotion

CATALIN

DeWald 561

Namco 601

Sentinel 177U

Motorola 52C

Emerson 400

Stewart Warner 9014E

Garod 1B551

Emerson 375

Fada 711

Emerson 190

Crosley G-1465

Motorola 50XC

Emerson 185

Emerson 197

Emerson 212

Emerson 214

Emerson 106

Emerson 570

Emerson 615B

Zenith 430

Silvertone 4666

Emerson 148

Emerson 229

Emerson 334

Ingraham Cabinets II

Emerson 316

Emerson 177

Emerson 418

Stewart-Warner 07-514H

Emerson 170

Silvertone 6120

Stromberg-Carlson 1000J

Emerson 141

Emerson 440

Emerson 350AW

Emerson 49

Firestone S-7403-3

Wood Cabinets

Halson 'Jefferson'

Emerson 321

Wurlitzer Lyric SW88

Garod 830

GE H-503

Gilfillan

Detrola 2811

Kadette

Automatic Tom Thumb

Majestic 669

Silvertone 47

Stromberg-Carlson 130U

CA 1949 REARING HORSE
POT METAL, PLATED METAL BASE
$350

CA 49 TWO REARING HORSES
POT METAL, PLATED METAL BASE
$400

Z477 CA 1949 LADY GODIVA
POT METAL STATUE, PLATED METAL BASE
$450

CA 1949 HULA GIRL
POT METAL, PLATED METAL, GYRATING WAISTE
$550

471 CA1935
2-TONE WOOD
$70

571 CA1935
WOOD
$75

572 CA1935
WOOD
$115

612 CA1935
2-TONE WOOD
$125

812 CA1935
WOOD
$120

'JR.' CA1934
WOOD
$225

'STANDARD 5' CA1934
WOOD
$90

52 CA1934
2-TONE WOOD, MANTLE
$75

100 CA1934
2-TONE WOOD CATHEDRAL
$150

121 CA1935
WOOD, SHOULDERED TOMBSTONE
$175

128 CA1935
WOOD, SQUAT TOMBSTONE
$110

132 CA1932
2-TONE WOOD FACE, METAL CABINET
$225

134 CA1935
WOOD CABINET, BLACK LACQUER TRIM
$125

138 CA1935
SKYSCRAPER-STYLED WOOD TOMBSTONE
$350

147 CA1935
WOOD, SQUAT TOMBSTONE
$110

152 CA1934
BAKELITE, SKYSCRAPER STYLED TOMBSTONE
$1500+

174 CA1935
WOOD, TOMBSTONE
$90

317 CA1935
2-TONE WOOD, TOMBSTONE
$125

322 CA1935
WOOD, TOMBSTONE
$90

4L-28 CA1959
COLORED PLASTIC
$55

5A-32 CA1952
BAKELITE CABINET, METAL TRIM
$50

5S-21 CA1952
BAKELITE
$35

5X-21 CA1952
BAKELITE CLOCK-RADIO
$45

5Z-22 CA1952
BAKELITE
$40

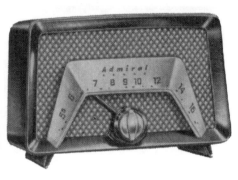

6C-22 CA1952
BAKELITE WITH CHROME TRIM
$45

6T-01 CA1946
BAKELITE
$35

6T-02 CA1946
BAKELITE
$35

6T-04 CA1946
WOOD
$25

6T-O5 CA1946
WOOD
$35

6T-O6 CA1946
WOOD
$45

6T-O7 CA1946
WOOD
$30

6T-11 CA1946
WOOD
$40

12-B5 CA1940
BROWN, BLACK BAKELITE
$50

13-C5 CA1940
IVORY PLASKON, BLACK TRIM
$75

15-B5 CA1940
BAKELITE
$110

17-B5 CA1940
WOOD
$40

18-B5 CA1940
WOOD
$45

20-A6 CA1940
BAKELITE
$110

22-A6 CA1940
WOOD
$40

25-Q5 CA1940
WOOD
$60

43-B4 CA1941
WOOD
$25

51-J35 CA1941
BAKELITE
$40

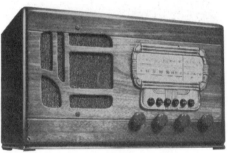

102-6B CA1938
WOOD
$65

113-5A CA1938
BAKELITE $150
PLASKON $200

125-5E 'JUNIOR' CA1938
BAKELITE $75, IVORY PLASKON $125,
COLORED PLASKON $300+

129-5F CA1939
WOOD, BLACK LACQUER TRIM
$115

133-7G CA1939
WOOD
$65

134-8A CA1939
WOOD
$70

141-4A CA1939
WOOD
$30

148-6K CA1939
WOOD
$60

158-5J CA1939
BAKELITE $65, IVORYPLASKON $115,
BEETLE PLASKON $300+

161-5L CA1939
BAKELITE $50, IVORYPLASKON $65,
BEETLE PLASKON $225

163-5L CA1939
BAKELITE $70, IVORYPLASKON $95,
BEETLE PLASKON $250+

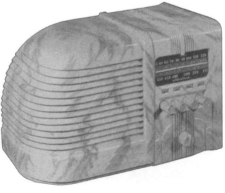

168-5D CA1939
BAKELITE $125, IVORYPLASKON $175,
BEETLE PLASKON $300+

169-5D CA1939
WOOD
$75

305-7C CA1940
WOOD
$60

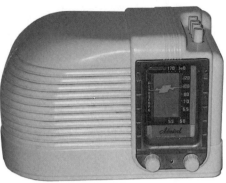

372-5R
BAKELITE $125
IVORY PLASKON $175

396-6M CA1939
BAKELITE $125, IVORYPLASKON $175,
BEETLE PLASKON $300+

399-6M CA1940
WOOD
$65

512-6C CA1938
BAKELITE $65
IVORY PLASKON $90

512-6D CA1938
WOOD
$65

516-4B CA1939
BAKELITE $40
IVORY PLASKON $60

547-6G CA1939
WOOD
$70

920-6Q CA1938
WOOD
$95

960-8K CA1938
WOOD WITH BLACK LACQUER TRIM
$110

975-6W CA1938
WOOD
$50

980-5X CA1938
WOOD
$60

4202-B6 CA1941
BEETLE PLASTIC
$175

A-126 CA1937
WOOD
$125

L-767 CA1936
WOOD
$135

M-169 CA1936
WOOD
$85

Y-300 CA1957
UREA, VARIOUS COLORS
$40

Z-344 CA1937
WOOD
$80

19-A66W CA1936
WOOD
$95

251 CA1935
2-TONE WOOD
$120

400 CA1934
METAL
$125

CA1938
WOOD
$95

J CA1937
WOOD
$110

(BY CLIMAX) CA1936
WOOD
$300

22 CA1939
BAKELITE $75
IVORY PLASKON $125

66 CA1934
BAKELITE $2,000+
COLORED PLASKON $2,000 - $10,000+

72 CA1938
WOOD
$110

100 CA1934
WOOD
$125

911 CA1939
WOOD
$60

871 CA1937
BAKELITE $500+
IVORY $1000+, COLORS $1500+

1001 CA1939
WOOD
$75

1001 CA1939
WOOD WITH ALUMINUM GRILLE
$300+

A-511 CA1950
BAKELITE
$50

A-520 CA1950
BAKELITE
$65

CA1934
CHROME FACE, WOOD CABINET
$400+

DORSET CA1932
WOOD
$275

'METEOR' CA1936
WOOD
$80

'MULTI-MU' CA1933
WOOD
$225

'STANTON' CA1932
WOOD
$250

3010 CA1938
BAKELITE
$350

3012 CA1938
IVORY PLASKON
$500+

4010 CA1938
BAKELITE
$350

4012 CA1938
BAKELITE
$350

A-10 CA1938
BAKELITE
$75

A-12 CA1938
BAKELITE
$100

A-3012 CA1938
BEETLE PLASTIC
$750+

14BR-526A CA1940
BAKELITE
$90

14BR-531A CA1940
WOOD, IVORY LACQUER FINISH
$150

62-164 CA1935
WOOD
$90

62-169 CA1936
WOOD
$75

62-177 CA1936
WOOD, BLACK LACQUER TRIM
$135

62-181 CA1936
WOOD
$120

62-196 CA1936
WOOD WITH BLACK LACQUER TRIM
$120

62-198 CA1936
WOOD
$110

62-211 CA1936
WOOD WITH BLACK LACQUER TRIM
$90

62-217 CA1936
WOOD
$175

62-230 CA1936
WOOD WITH BLACK LACQUER TRIM
$60

62-304 'BULLET' CA1937
2-TONE WOOD
$150

62-600 CA1938
WOOD
$90

93BR-715A CA1939
WOOD
$125

GSL1561-A
PAINTED BAKELITE CLOCK-RADIO
$35

P362C-2568 CA1941
WOOD
$70

P362C-2673 CA1941
WOOD
$75

P462C-610 CA1941
BAKELITE
$85

P462C-612 CA1941
WOOD
$50

P462C-729 CA1941
WOOD
$110

P462C-731 CA1941
WOOD
$40

4 CA1935
WOOD
$75

5 CA1935
WOOD
$125

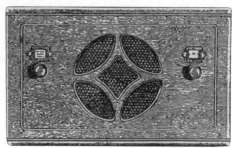

5A CA1932
WOOD
$65

200A CA1932
WOOD-TREASURE CHEST
$300

200C 'BOMBAY' CA1932
WOOD-TREASURE CHEST
$400

236A CA1932
WOOD
$90

355 CA1934
WOOD
$125

360T/370T CA1934
WOOD
$150

402 CA1935
WOOD
$90

42O CA1935
WOOD WITH METAL INLAY
$125

43OT CA1936
WOOD
$150

44OT CA1935
WOOD WITH BLACK LACQUER TRIM
$225

46OA CA1935
WOOD WITH METAL INLAY
$225

46OB CA1935
WOOD
$125

47OU CA1935
WOOD WITH BLACK LACQUER TRIM
$165

5OO CA1933
WOOD WITH MARQUETRY INLAY
$100

5O1 CA1934
WOOD WITH MARQUETRY INLAY
$125

5O5 CA1936
WOOD
$50

510 CA1936
Wood
$115

515 CA1937
Wood
$40

575F CA1936
Wood
$125

585Y CA1936
Wood with Marquetry Inlay
$140

604 CA1937
Wood
$45

605 CA1937
Wood
$45

610 CA1937
Wood
$40

640 CA1937
Wood
$45

660T CA1937
Wood
$60

41 CA1946
WOOD
$110

51 CA1936
WOOD
$125

61 CA1936
WOOD
$150

89 CA1939
WOOD
$70

160 CA1948
BAKELITE
$30

241P CA1949
PLASTIC
$65

242T CA1949
PAINTED METAL
$75

356T CA1950
BAKELITE
$55

360TFM CA1949
BAKELITE
$35

446P CA1950
PLASTIC
$65

450T CA1953
PAINTED BAKELITE
$60

451T CA1953
PAINTED BAKELITE, FOIL LIT DIAL
$65

460T CA1953
PLASTIC
$25

518 CA1938
WOOD
$65

518A 'PHANTOM GIRL' CA1938
IVORY LACQUER WOOD
$225

518DW 'PHANTOM PAL' CA1938
IVORY LACQUER & NATURAL WOOD
$225

544 CA1946
PAINTED BAKELITE
$70

551 CA1953
WOOD
$40

552 CA1947
PAINTED BAKELITE
$40

553T CA1951
BAKELITE
$50

568DW 'PHANTOM ACE' CA1938
IVORY LACQUER & NATURAL WALNUT
$250

578B CA1938
WOOD
$50

618 'PHANTOM MAID' CA1938
WOOD
$75

618 'PHANTOM SENIOR' CA1938
WOOD
$65

622A CA1941
PAINTED BAKELITE
$85

651T CA1951
PAINTED BAKELITE
$60

657T CA1951
PAINTED BAKELITE CLOCK-RADIO
$50

664 CA1946
BAKELITE
$40

741T CA1954
PAINTED BAKELITE
$80

753T CA1954
BAKELITE
$65

758T CA1954
PLASTIC CLOCK-RADIO
$35

760T CA1954
PLASTIC
$25

952P1 CA1955
COLORED PLASTIC
$55

80 CA1932
WOOD
$550

82 CA1932
WOOD
$550

84 CA1931
WOOD
$350

90 CA1931
WOOD
$550

93 (SW CONVERTER) CA1932
WOOD (SHOWN BENEATH AK90 CATHEDRAL)
$225

145 CA1934
WOOD WITH MARQUETRY INLAY
$175

165 CA1933
WOOD WITH BLACK LACQUERT INSERT GRILLE
$325

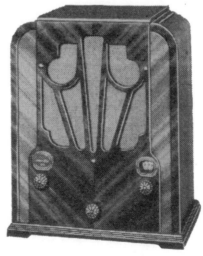

184 CA1936
WOOD
$255

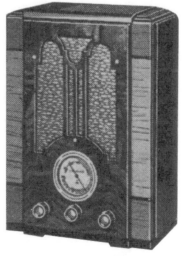

225 CA1936
WOOD
$175

228 CA1933
WOOD
$525

237 CA1935
WOOD
$325

246 CA1933
WOOD
$450

337 CA1936
WOOD
$275

356 CA1935
WOOD
$325

387 CA1933
WOOD
$475

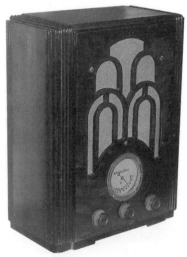

451Q CA1935
WOOD
$150

465 CA1934
WOOD
$375

545 CA1935
WOOD
$175

558 CA1932
WOOD
$550

567 CA1932
WOOD
$475

627 CA1932
WOOD
$475

725A CA1935
WOOD
$175

735 CA1935
WOOD
$325

856 CA1936
WOOD
$185

Audiola

OO5W CA1933
WOOD WITH MARQUETRY INLAY
$90

1O CA1931
WOOD
$225

517 CA1933
REPWOOD FACE, WOOD CABINET
$275

811 CA1933
WOOD
$150

MIDGET CA1931
WOOD
$250

S-7 CA1933
METAL CABINET
$80

CA1933
WOOD WITH BLONDE & BLACK LACQUER TRIM
$350

CA1937
BAKELITE
$350

CA1939
BAKELITE
$75

440 CA1940
BAKELITE
$60

933 (TOM THUMB) CA1939
METAL/VARIOUS COLORS $1000+
CATALIN/VARIOUS COLORS $2500+

3749 CA1940
WOOD WITH CHROME INSERT GRILLE BARS
$135

ATLAS CA1937
WOOD WITH BLACK LACQUER DETAIL
$125

MIDGET CA1931
WOOD
$150

TOM THUMB PEE WEE CA1939
WOOD
$450

6D15 Skyrover CA1946
Ivory Painted Wood
$70

71A CA1933
Wood
$275

425 Freshman CA1933
Metal Cabinet
$125

522 CA1936
Wood
$80

545 CA1933
Wood with Marquetry Inlay
$110

571 Clockette CA1940
Wood Clock-Radio
$90

571 Skyrover CA1940
Wood
$75

575 'Coat & Vest' CA1936
Wood
$225

582 CA1938
Wood with Tenite Trim
$65

585 CA1935
WOOD
$150

585D FRESHMAN CA1935
WOOD
$185

601 CA1936
WOOD, 2-TONE LACQUER
$85

629 CA1940
WOOD
$50

637 FRESHMAN CA1940
WOOD
$55

675 CA1937
WOOD
$190

686 CA1936
WOOD
$70

775T CA1935
WOOD
$225

777 CA1936
WOOD
$165

778A CA1936
Wood
$85

791B CA1939
Wood
$60

PERSONAL CA1933
METAL CABINET
$125

T-30 CA1952
PLASTIC
$25

T-522 CA1953
PLASTIC
$45

TC-20 CA1953
PLASTIC CLOCK-RADIO
$45

TC-500 CA1953
PLASTIC CLOCK-RADIO
$25

MIDGET 6 CA1932
WOOD
$175

MIDGET 8 CA1932
WOOD
$200

40 'JUNIOR' CA1932
WOOD
$200

60 'JUNIOR' CA1931
WOOD
$175

61 CA1931
WOOD
$190

70 CA1931
WOOD
$225

70X CA1938
WOOD WITH BLACK LACQUER TRIM
$75

80 CA1931
WOOD
$225

85 CA1932
WOOD
$225

420 CA1933
WOOD WITH INLAY
$100

422 CA1933
WOOD MINI-CATHEDRAL
$145

450 CA1933
WOOD WITH MARQUETRY INLAY
$115

470 CA1934
WOOD
$170

471 CA1934
WOOD $150
WOOD WITH CHROME GRILLE $350+

490 CA1934
WOOD WITH MARQUETRY INLAY
$100

500 CA1934
WOOD
$190

511 CA1934
WOOD WITH CHROME GRILLE
$350+

691 CA1938
WOOD WITH BLACK LACQUER TRIM
$95

770 CA1938
WOOD WITH BLACK LACQUER TRIM
$55

11802 CA1940
BAKELITE WITH TENITE INSERT GRILLE
$75

TC53 CA1935
WOOD WITH BLACK LACQUER TRIM
$150

CA1934
WOOD WITH CHROME GRILLE, BLACK LACQUER TRIM
$250+

CA1936
2-Tone Wood
$175

CA1936
2-Tone Wood Mini-Tombstone
$225

CA1937
Wood
$500+

CA1937
Wood
$175

Emerald (1st Version) CA1937
Wood
$750+

CA1937
Wood
$325

90, 91 'Oval' CA1936
Wood
$500+

4 CA1932
Wood
$135

8 CA1932
Wood
$160

39 CA1931
2-Tone Wood
$250

44 CA1931
Wood
$225

250 CA1933
Wood with Marquetry Inlay
$200

279 CA1933
2-Tone Wood, Marquetry, Insert Grille
$225

625 CA1933
2-Tone Wood
$190

650 CA1934
Wood, Mini-Tombstone, Brass Inlay
$350+

654 CA1934
Wood with Marquetry Inlay
$70

655 CA1935
2-TONE WOOD
$125

656 CA1935
2-TONE WOOD
$150

657 CA1935
2-TONE WOOD
$125

658 CA1935
2-TONE WOOD, INSERT GRILLE
$200

MIDGET CA1930
WOOD
$225

R6764 CA1931
WOOD
$150

T397 CA1933
WOOD
$250

Coronado

CA1933
Wood
$225

7P CA1937
Wood
$65

31B CA1935
2-TONE WOOD
$115

115 'ROCKET' CA1953
BAKELITE
$30

778 CA1936
WOOD
$75

806A CA1938
WOOD
$50

1040A CA1940
WOOD
$35

1-N 'LITLFELLA' CA1932
WOOD
$150

6H2 CA1935
WOOD WITH CHROME BEZEL
$200

9-103 CA1949
PAINTED BAKELITE
$60

9-113 CA1949
BAKELITE
$25

9-118 CA1949
BAKELITE
$60

9-119 CA1949
BAKELITE
$75

10-127 CA1950
BAKELITE
$90

10-304 'PLAYTIME' CA1950
BAKELITE
$70

11-3046 CA1951
PAINTED BAKELITE
$65

32 CA1934
WOOD
$220

36AM CA1941
2-TONE WOOD
$25

47H9 CA1934
WOOD WITH CHROME BEZEL
$150

48 'ELF' CA1931
REPWOOD
$350

61 CA1936
WOOD
$125

54 'NEW BUDDY' CA1931
REPWOOD
$250

54G 'NEW BUDDY BOY' CA1931
REPWOOD
$325

56FP CA1946
WOOD
$70

58 'BUDDY BOY' CA1931
REPWOOD
$350

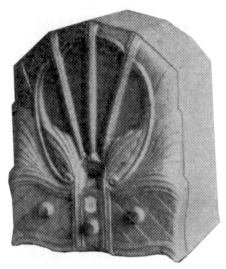

59 'SHOWBOY' CA1931
REPWOOD
$250

61AF CA1935
WOOD
$135

66TA CA1946
BAKELITE
$75

124 'PLAYTIME' CA1931
2-TONE WOOD
$165

148 'LIBRARY DELUXE' CA1932
REPWOOD
$225

148 'FIVER' CA1931
WOOD
$150

163 'TRAVO DELUXE' CA1933
BLACK METAL WITH CHROME ESCUTCHEONS
$150

435 CA1936
WOOD
$90

516 CA1935
WOOD
$90

517-7H CA1939
2-Tone Wood
$60

547 CA1937
2-Tone Wood
$150

614EH CA1935
Wood with Chrome Bezel
$150

629 CA1937
Wood
$95

634 CA1937
Wood
$115

635 CA1935
Wood
$85

638B CA1939
Wood
$125

718A CA1939
Wood
$135

744 CA1937
Wood
$115

745 CA1937
WOOD
$110

814 CA1935
WOOD WITH CHROME BEZEL
$145

817 CA1938
WOOD
$55

5628B CA1939
PAINTED BAKELITE
$85

B-667A CA1938
WOOD
$90

BATTERY 8 CA1934
WOOD
$120

BATTERY FIVER CA1935
WOOD
$60

OLYMPIA 6 CA1936
WOOD WITH CHROME BEZEL
$110

C-526 CA1937
WOOD
$95

C-629 CA1937
Wood
$80

'CENTURION' CA1936
Wood (10-tube)
$250

'CLIPPER' CA1936
Wood
$140

'COMPANION' CA1933
Wood
$125

'MERRIMAC 8' CA1935
Wood with Brass Escutcheion
$150

'CROSLEY A.F.M.' CA1936
Wood
$110

'DUAL 4' CA1934
Wood
$150

'DUAL 5' CA1934
Wood
$160

'DUAL 7' CA1934
Wood
$225

'DUAL 10' CA1934
WOOD
$300

'DUAL 12' CA1934
WOOD
$350

'DUAL 5 SHERATON' CA1934
WOOD
$325

'DUAL CASA' CA1934
WOOD
$115

''DUAL COMPANION' CA1934
WOOD
$125

'DUAL TRAVETTE' CA1934
BLACK METAL WITH CHROME ESCUTCHEONS
$200

E-10 CA1953
PAINTED BAKELITE
$110

'FIVER COMPACT' CA1938
WOOD
$40

'FIVER' CA1933
WOOD
$170

'FIVER' CA1937
2-TONE WOOD
$125

'FIVER' CA1936
WOOD
$70

'GRADUATE' CA1954
COLORED PLASTIC
$75

'JEWEL CASE' CA1933
WOOD
$300

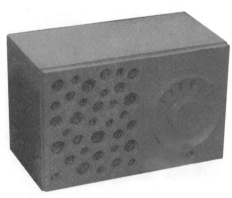

JT3 CA1955
PAINTED BAKELITE, PLASTIC
$100

'LEADER D' CA1933
WOOD
$160

'MAYOR' CA1932
WOOD
$225

'NEW TRAVO' CA1933
WOOD
$90

'PLAYBOY' CA1932
WOOD
$170

'REPOSE JR.' CA1934
WOOD
$140

'SEXTET' CA1933
WOOD
$165

'SUPER 8' CA1938
WOOD
$75

SW CONVERTER CA1934
WOOD
$110

'TENSTRIKE' CA1932
WOOD
$225

'TWELVE' CA1932
WOOD
$350

'TYNAMITE' CA1932
2-TONE WOOD
$175

'MONITOR 8' CA1936
WOOD
$125

'WIGIT' CA1932
REPWOOD
$375

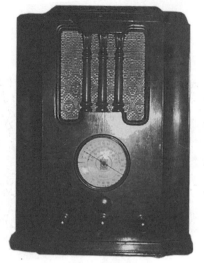

1107 CA1935
Wood
$110

R1115 CA1937
Wood
$65

R1232 CA1947
Wood
$50

R1234 CA1947
Painted Bakelite
$20

R1236 CA1947
Painted Bakelite
$25

R1238 CA1947
Wood
$30

R1238 CA1948
Plastic, Bakelite
$50

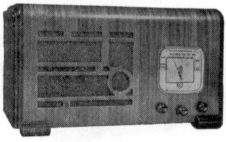

R2055 CA1938
Wood
$45

R3210 CA1938
Wood
$35

CA1932
2-Tone Wood
$150

5D CA1934
Wood with Black & Chrome Lacquer Trim
$250

120 CA1937
Wood
$300

134XA1 CA1937
Wood with Beetle Dial Bezel
$70

139 CA1937
Wood with Shadowed Finish
$175

139EA CA1937
Black Lacquer Wood with Chrome Trim
$400

140A CA1937
Wood
$75

146EA1 CA1937
Wood
$90

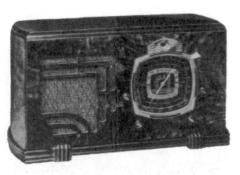

147EA1 CA1937
Wood
$125

148EA CA1937
WOOD
$150

159A CA1937
WOOD
$35

177A CA1937
WOOD
$30

178A CA1937
WOOD
$45

182A CA1937
WOOD WITH BEETLE DIAL BEZEL
$70

193E CA1938
WOOD, MOTOR DIAL
$350

209EA CA1938
WOOD
$60

240 CA1938
WOOD
$225

274 CA1939
COLORED PLASKON WITH TENITE GRILLE
$750+

302 CA1940
WOOD WITH MARQUETRY, CLOCK-RADIO
$350

304 CA1939
WOOD WITH TENITE GRILLE
$45

319 CA1940
WOOD
$35

321 CA1939
WOOD WITH TENITE GRILLE
$45

429 CA1941
WOOD
$40

571 CA1947
WOOD WITH METAL GRILLE
$40

2811 CA1939
WOOD WITH MARQUETRY INLAY
$350+

3041 CA1939
WOOD
$35

3101 CA1939
WOOD
$35

3201 CA1939
WOOD
$40

3202 CA1939
WOOD
$35

PEE WEE CONVERTER CA1940
BAKELITE
$250+

SHORT WAVE CONVERTER CA1933
WOOD
$175

1 CA1933
WOOD
$115

401B 'BANTAM' CA1940
BAKELITE WITH PLASKON TRIM
$150

407 CA1940
BAKELITE WITH PLASKON TRIM
$250

414 CA1933
WOOD
$135

425 CA1934
WOOD
$125

440 CA1935
REPWOOD
$250

441 'MIGHTY MITE JR.' CA1935
2-TONE WOOD
$110

442 'MIGHTY MITE SR.' CA1935
2-TONE WOOD
$175

501B CA1935
2-TONE WOOD, BLACK LACQUER TRIM
$300

503A CA1935
WOOD
$125

507 CA1935
2-TONE WOOD
$95

520 CA1937
2-TONE WOOD
$100

521 CA1937
2-TONE WOOD
$140

522 CA1937
WOOD
$60

524 CA1931
2-TONE WOOD., INSERT GRILLE
$225

524 CA1931
WOOD WITH INSERT GRILLE
$240

530 CA1938
PAINTED BAKELITE
$45

551 CA1933
WOOD
$135

561 CA1939
CATALIN
$700 - $2000+

600A CA1935
WOOD
$90

603 CA1935
WOOD
$90

610SA CA1936
WOOD
$120

619 CA1937
WOOD
$125

645 CA1939
WOOD
$60

648 CA1939
WOOD
$60

701 CA1938
WOOD
$60

740 CA1933
2-TONE WOOD
$175

801 CA1934
WOOD WITH INSERT GRILLW
$160

802 CA1935
WOOD WITH INSERT GRILLE
$175

1200 CA1938
WOOD
$40

A503R CA1948
WOOD
$30

A514 'BANTAM' CA1948
IVORY PLASKON
$125

B504 CA1949
2-COLOR PLASTIC
$65

B506 CA1948
IVORY PLASKON
$70

B622FM CA1948
WOOD
$25

C800 CA1949
BAKELITE
$20

60 CA1932
Wood with Repwood Grille, Escutcheon
$350

80 CA1932
Wood with Repwood Grille, Escutcheon
$350

S-3 CA1933
2-Tone Wood, Insert Grille
$200

S-5 CA1932
Wood with Repwood Trim
$250

CA1935
Wood
$125

EC113 (Hallicrafters) CA1947
Wood
$125 (Hallicarfters 1st Home Model)

26 CA1934
WOOD
$70

31-LO4 CA1949
COLORED PLASTIC CLOCK-RADIO
$25

33AW CA1934
WOOD WITH BLACK LACQUER & ALUMINUM TRIM
$300+

34-F7 CA1935
WOOD, INGRAHAM CABINET
$100

35 'SHERATON' CA1933
WOOD, INGRAHAM CABINET
$300

36 CA1935
WOOD
$75

59 CA1935
WOOD, INGRAHAM CABINET
$90

60 CA1935
WOOD, INGRAHAM CABINET
$750+

71 CA1935
WOOD, INGRAHAM CABINET
$225

104 CA1936
WOOD, INGRAHAM CABINET
$200

105 CA1936
WOOD, INGRAHAM CABINET
$750+

106-A CA1936
WOOD, INGRAHAM CABINET
$200

109 CA1936
BAKELITE
$150

117 'BUTTERFLY' CA1936
WOOD, INLAID PLASTIC, INGRAHAM CABINET
$100

117-A 'BUTTERFLY' CA1936
WOOD, INLAID PLASTIC, INGRAHAM CABINET
$100

118 CA1936
REPWOOD
$125

121 CA1936
WOOD, BRASS GRILLE BARS, INGRAHAM CABINET
$65

122 CA1936
WOOD, BLACK LACQUER TRIM, INGRAHAM CABINET
$45

Emerson

123 CA1936
WOOD, INGRAHAM CABINET
$45

140 CA1937
WOOD, INGRAHAM CABINET
$250

152 CA1937
WOOD, INGRAHAM CABINET
$115

171 CA1937
WOOD, INGRAHAM CABINET
$125

172 CA1937
WOOD, INGRAHAM CABINET
$400

177 CA1937
WOOD, INGRAHAM CABINET
$750+

179 CA1937
WOOD, INGRAHAM CABINET
$50

185 CA1937
WOOD, INGRAHAM CABINET
$400

187 'PAGODA' CA1937
REPWOOD, WHITE WITH GOLD ACCENTS
$350

192 CA1938
WOOD, INGRAHAM CABINET
$30

204 CA1938
WOOD, INGRAHAM CABINET
$50

206 CA1938
BAKELITE
$45

207 CA1938
WOOD, INGRAHAM CABINET
$50

209 CA1938
WOOD, INGRAHAM CABINET
$60

213 CA1938
WOOD, INGRAHAM CABINET
$350

215 CA1938
WOOD, INGRAHAM CABINET
$45

228 CA1938 (SAKNOFFSKY DES.)
WOOD, INGRAHAM CABINET, CONICAL DIAL

234 CA1938
WOOD, INGRAHAM CABINET
$65

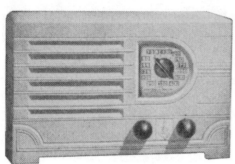

253 CA1939
PRESSBOARD
$100

261 CA1939
WOOD, INLAID COLORED PLASTIC, INGRAHAM CAB.
$135

264 CA1939
PRESSBOARD
$90

268 CA1940
BAKELITE
$40

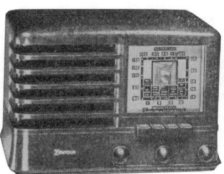

269 CA1940
BAKELITE
$50

271 CA1940
WOOD, INGRAHAM CABINET
$90

276 CA1940
WOOD, INGRAHAM CABINET
$40

279 CA1940
BAKELITE $150
COLORED PLASKON $350+

282 CA1940
WOOD, INGRAHAM CABINET
$225

285 CA1940
WOOD, INGRAHAM CABINET
$25

286 CA1940
WOOD, INGRAHAM CABINET
$35

287 CA1940
WOOD, INGRAHAM CABINET
$45

296 CA1940
WOOD, INGRAHAM CABINET
$35

300 'JEWEL CHEST' CA1933
WOOD, BURL WITH BLACK LACQUER & CHROME TRIM
$350

301 CA1940
BAKELITE
$35

305 CA1940
WOOD, INGRAHAM CABINET
$150

318 CA1940
WOOD, INGRAHAM CABINET
$120

321 'CHINESE CHEST' CA1934
WOOD, HANDPAINTED LACQUERS
$400+

331 CA1940
WOOD, INGRAHAM CABINET
$35

332 CA1940
WOOD, INGRAHAM CABINET
$40

336 CA1941
BAKELITE
$35

350AW CA1934
WOOD, INGRAHAM CABINET
$350+

354 CA1941
WOOD, INGRAHAM CABINET
$45

365 CA1941
WOOD, INGRAHAM CABINET
$125

375 CA1934
WOOD, INGRAHAM CABINET, TAMBOR DOORS
$350

375 CA1941
CATALIN
$1000+

376 CA1941
WOOD, INGRAHAM CABINET
$175

405 'Patriot Chest' ca1942
Wood, Ingraham Cabinet
$150

413 ca1942
Bakelite
$40

414 'Big Six' ca1942
Bakelite
$35

415 ca1934
Bakelite
$175

416 ca1934
Wood, Ingraham Cabinet
$150

421 ca1942
Bakelite
$30

425 ca1942
Wood, Ingraham Cabinet
$35

426 ca1942
Bakelite
$45

439 ca1942
Wood, Ingraham Cabinet
$40

441 CA1942
BAKELITE
$25

455 CA1942
WOOD, INGRAHAM CABINET
$250

459 CA1933
2-TONE WOOD
$75

460 CA1942
WOOD
$25

467 CA1942
WOOD, INGRAHAM CABINET
$40

509 CA1946
BAKELITE
$25

516 CA1947
BAKELITE
$25

520 CA1931
WOOD
$150

540 CA1947
BAKELITE $125
COLORED PLASKON $175-$350

556 CA1933
2-TONE WOOD
$225

556 CA1950
WOOD
$20

557 CA1933
2-TONE WOOD
$145

559 'TREASURE CHEST' CA1933
WOOD
$350

570 'MOMENTO' CA1949
BLACK BAKELITE, GOLD PLASTIC TRIM
$110

577 CA1950
WOOD
$25

581 CA1950
IVORY PLASKON
$45

599 CA1949
BAKELITE
$20

602C CA1949 (LOEWEY DES.)
COLORED PLASTIC WITH GOLD TRIM
$75

613 CA1949
COLORED PLASTIC
$60

641 CA1951
BAKELITE
$25

646 CA1951
COLORED PLASTIC
$45

652 CA1951
COLORED PLASTIC
$35

653 CA1951
BAKELITE
$25

659 CA1951
BAKELITE
$25

671 CA1951
BAKELITE CLOCK-RADIO
$40

705 CA1952
COLORED PLASTIC, GOLD PLASTIC TRIM
$50

706 CA1952
BAKELITE $70
COLORED UREA $100-$175

707 'Sunburst' CA1953
BAKELITE $90
COLORED UREA $125-$200

708 CA1954
PLASTIC
$20

729 CA1954
COLORED UREA WITH REVERSE-PAINTED PLASTIC GRILLE
$75

755M CA1933
WOOD
$180

778 CA1955
COLORED PLASTIC
$65

788 CA1955
PLASTIC CLOCK-RADIO
$40

808 CA1955
COLORED UREA, PLASTIC
$60

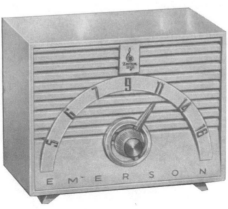

811 CA1955
PLASTIC
$25

812 CA1955
COLORED UREA
$60

813 CA1955
BAKELITE
$30

816 CA1955
COLORED PLASTIC CLOCK-RADIO
$40

825 CA1955
COLORED PLASTIC, CLEAR LUCITE BASE
$45

826 CA1955
COLORED UREA WITH CLEAR LUCITE TRIM
$125

832 CA1956
PLASTIC
$15

846 CA1956
COLORED UREA WITH CLEAR LUCITE TRIM
$65

851 CA1956
2-TONE COLORED PLASTIC
$65

852 CA1956
COLORED PLASTIC
$45

853 CA1956
COLORED PLASTIC
$35

876 CA1957
COLORED PLASTIC CLOCK-RADIO
$25

881 CA1957
COLORED PLASTIC CLOCK-RADIO
$25

883 CA1957
COLORED PLASTIC CLOCK-RADIO
$25

915 CA1958
COLORED PLASTIC
$35

'LIBRARY' CA1947
PRESSBOARD
$150

'ENFANT' CA1930
REPWOOD
$500+

2D5 (ANDREA) CA1938
WOOD
$65

2D8 (ANDREA) CA1938
WOOD
$75

2E6 (ANDREA) CA1939
2-TONE WOOD
$110

2E8 (ANDREA) CA1939
WOOD
$120

4E11 (ANDREA) CA1939
2-TONE WOOD
$80

12E6 (ANDREA) CA1939
2-TONE WOOD
$90

2OT CA1939
WOOD
$55

43 CA1931
WOOD WITH REPWOOD FACE
$300

44 CA1939
BAKELITE
$75

51 CA1931
WOOD
$225

55 CA1933
WOOD WITH BLACK LACQUER TRIM
$150

73 CA1932
2-TONE WOOD
$90

106 'SUPER FADALETTE' CA1933
WOOD WITH INLAID MARQUETRY
$120

110AM CA1935
WOOD WITH IVORY & BLACK LACQUER FINISH
$200

130 CA1935
WOOD WITH INSERT GRILLE
$145

135 CA1935
WOOD
$125

140 CA1935
WOOD
$155

141 CA1934
WOOD WITH INLAID MARQUETRY & BLACK LACQUER
$175

155 CA1935
WOOD
$110

163T CA1936
WOOD
$150

172 CA1937
WOOD
$125

190T CA1936
2-TONE WOOD
$125

209V CA1941
IVORY PLASKON
$90

260T CA1937
WOOD
$225

211T CA1937
2-TONE WOOD
$1000+ (11-TUBE); $750+ (9 TUBE)

212T CA1937
WOOD
$1500+ (16-TUBE); $1400+ (12 TUBE)

212T (LATER VERSION) CA1938
WOOD
$1200+

260V CA1937
BLACK BAKELITE WITH CHROME TRIM
$350+

262T CA1937
WOOD
$225

265T CA1937
WOOD
$150

270T CA1937
WOOD
$225 (7 TUBES)

280T CA1937
WOOD
$250 (8 TUBES)

350V CA1937
IVORY PLASKON
$250

351K CA1938
WOOD
$65

358 CA1938
WOOD
$75

450T CA1937
WOOD WITH BRASS TRIM
$80

450W CA1938
BAKELITE WITH BRASS TRIM
$275

451T CA1938
WOOD
$60

605 CA1947
IVORY PLASKON
$80

609 CA1946
BAKELITE
$55

626 (ANDREA) CA1939
WOOD
$65

629 (ANDREA) CA1939
WOOD
$125

659 'TEMPLE' CA1947
CATALIN
$500-$3000+

700 CA1947
CATALIN
$500-$2500+

711 CA1947
CATALIN WITH PLASTIC TRIM
$500 - $2000+

777 CA1947
IVORY PLASKON
$90

855 CA1949
PLASTIC
$90

1000 CA1946
CATALIN
$500-$3000

1001 CA1946
WOOD
$35

1002 CA1946
WOOD
$55

1452A CA1935
WOOD
$115

1462D CA1935
WOOD
$160

1470C CA1935
WOOD
$90

1582H CA1935
WOOD
$135

EL263 CA1937
WOOD WITH COPPER & CORK TRIM
$225

L56MA CA1940
CATALIN
$2000-$10,000

P111 CA1949
BAKELITE
$75

P38 CA1947
BAKELITE
$125

SilentRadio CA1948
BAKELITE
$150 (WITH SPEAKER)

SUPER FADALETTE CA1933
PRESSBOARD
$85

CA1935
WOOD WITH MARQUETRY INLAY & BLACK TRIM
$175

5AT1 CA1938
2-TONE WOOD
$75

58 CA1937
EXOTIC WOODS WITH BLACK LACQUER TRIM
$200

58T2 CA1937
WOOD WITH BLACK LACQUER TRIM
$150

72T3 CA1937
WOOD
$135

91T4 CA1937
WOOD
$175

6010 CA1935
EXOTIC WOODS
$225

7014 CA1935
WOOD WITH BLONDE & BLACK LACQUER DEATIL
$350

7117 CA1936
WOOD
$110

CA1936
WOOD
$110

AT10 CA1940
BAKELITE
$125

AT16 CA1940
WOOD WITH BLACK LACQUER TRIM
$150

AT21 CA1940
BEETLE PLASTIC
$250

AT22 CA1940
WOOD WITH BLACK LACQUER TRIM
$110

AT40 CA1940
WOOD WITH BLACK LACQUER TRIM
$20

AT50 CA1940
WOOD
$30

AT51 CA1940
EXOTIC WOODS
$50

AT52 CA1940
WOOD
$45

BT20 CA1940
BAKELITE
$70

BT22 CA1940
WOOD
$45

BT55 CA1941
WOOD
$45

BT61 CA1941
WOOD
$45

BT70 CA1941
WOOD
$30

BT71 CA1941
WOOD
$30

CT42 CA1941
WOOD
$40

GT55 CA1948
PAINTED BAKELITE
$25

GTO60 CA1948
BAKELITE WITH MARBELED PLASTIC GRILLE
$50

GTO64 CA1948
BAKELITE
$25

Firestone

4-A-21 CA1947
WOOD
$50

4-A-24 CA1941
2-TONE WOOD
$30

R-320 'WORLD'S FAIR' CA1939
PAINTED BAKELITE
$325

S-7403-3 CA1940
EXOTIC WOODS, INGRAHAM CABINET
$225

S-7403-4 CA1940
WOOD WITH INLAID PLASTIC, INGRAHAM CABINET
$125

S-7403-5 CA1940
WOOD, INGRAHAM CABINET
$125

S-7403-6 CA1940
WOOD, INLAID PLASTIC, INGRAHAM CABINET
$175

S-7426-6 CA1940
BEETLE PLASTIC
$225

'TIMETUNER' CA1939
WOOD CLOCK-RADIO
$125

CA1939
BAKELITE
$115

'RIBBON GRILLE' CA1939
PAINTED BAKELITE
$60

CA1950
RED PLASTIC
$30

30D CA1938
WOOD
$80

55 CA1933
WOOD WITH INLAID MARQUETRY
$100

55P CA1933
CANVAS BOARD
$50

56 CA1936
WOOD
$80

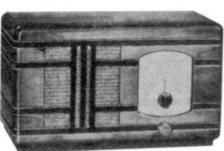

60 CA1937
WOOD
$75

62 CA1937
WOOD
$135

77 CA1933
WOOD
$165

96 CA1932
WOOD
$275

346 CA1934
CANVAS BOARD
$70

350 CA1934
WOOD
$175

360 CA1934
WOOD
$175

367 CA1935
WOOD
$175

469 CA1935
WOOD
$75

FM-40 CA1941
WOOD
$80

CA1933
WOOD
$200

CA1936
WOOD
$95

CA1936
WOOD
$85

JEWEL CHEST CA1933
WOOD
$325

4B1 'PIONEER' CA1941
2-TONE WOOD
$40

5A1 'ENSIGN' CA1941
IVORY PLASKON
$150

5A3 'AMBASSADOR' CA1941
2-TONE WOOD
$40

6B1 'SENATOR' CA1941
PAINTED BAKELITE
$90

6D1 'GOVERNOR' CA1941
WOOD
$45

6F1 'ESQUIRE' CA1941
2-TONE WOOD
$40

8AS1 'PRESIDENT' CA1941
WOOD
$60

25 CA1935
WOOD
$175

27 CA1935
WOOD
$135

830 CA1937
WOOD
$500+

D1443 'ENSIGN' CA1940
BAKELITE
$110

G35 CA1934
WOOD
$130

50 CA1948
BAKELITE CLOCK-RADIO
$25

60 CA1948
BAKELITE CLOCK-RADIO
$20

65 CA1949
PLASTIC CLOCK-RADIO
$25

67 CA1949
PAINTED BAKELITE
$30

123 CA1950
PLASTIC
$25

136 CA1950
PLASTIC
$40

145 CA1950
PLASTIC
$45

210 CA1948
BAKELITE
$25

218 CA1950
PLASTIC
$20

226 CA1950
PLASTIC
$30

356 CA1948
BAKELITE
$25

401 CA1950
PLASTIC CLOCK-RADIO
$25

410 'JUNIOR' CA1931
WOOD WITH REPWOOD GRILL
$150

419 CA1954
COLORED PLASTIC
$35

424 CA1954
BAKELITE
$45

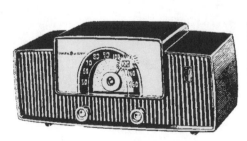

427 CA1954
COLORED PLASTIC
$45

436 CA1954
PLASTIC
$35

440 CA1954
BAKELITE
$45

472 CA1956
PLASTIC
$15

475 CA1956
COLORED PLASTIC
$25

546 CA1954
COLORED PLASTIC CLOCK-RADIO
$35

555 CA1954
COLORED PLASTIC CLOCK-RADIO
$25

560 CA1954
COLORED PLASTIC CLOCK-RADIO
$25

564 CA1954
COLORED PLASTIC CLOCK-RADIO
$30

612 CA1954
COLORED PLASTIC
$125

740 CA1941
WOOD
$50

860 'ATOMIC' CA1957
PLASTIC
$65

895 CA1956
COLORED PLASTIC CLOCK-RADIO
$15

900 CA1956
COLORED PLASTIC CLOCK-RADIO
$20

915 CA1956
COLORED PLASTIC CLOCK-RADIO
$45

916 CA1956
COLORED PLASTIC CLOCK-RADIO
$20

921 CA1956
COLORED PLASTIC CLOCK-RADIO
$35

A-51 CA1937
WOOD
$60

A-63 CA1936
WOOD
$90

A-64 CA1937
WOOD
$110

A-82 CA1936
WOOD
$150

B-81 CA1933
WOOD
$180

B-X CA1933
METAL
$85

C-70 CA1935
WOOD
$130

E-51 CA1936
WOOD
$70

E-52 CA1937
WOOD, PAINTED WOOD
$110

E-61 CA1936
WOOD
$85

E-62 CA1936
WOOD
$50

E-71 CA1936
WOOD
$115

E-72 CA1936
WOOD
$45

E-81 CA1936
WOOD
$85

E-91 CA1936
WOOD
$85

E-101 CA1936
WOOD
$125

F-53 CA1937
WOOD
$80

F-70 CA1937
WOOD
$55

FB-52 CA1939
WOOD
$125

FB-72 CA1939
WOOD
$110

FE-51 CA1937
2-TONE WOOD
$125

G-53 CA1939
WOOD
$60

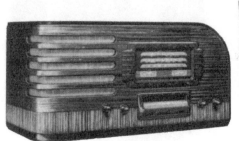

G-61 CA1939
WOOD
$60

G-65 CA1939
WOOD
$50

GD-41 CA1938
WOOD
$80

GD-52 CA1939
WOOD
$125

GD-62 CA1939
WOOD
$80

GD-63 CA1939
EXOTIC WOODS
$225

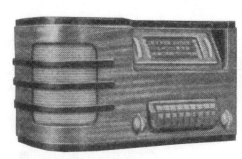

H-73 CA1940
WOOD WITH BLACK LACQUER GRILLE BARS
$90

H-400 'PEE WEE' CA1940
BAKELITE
$135

H-503 CA1939
WOOD
$150

H-51OU CA194O
BAKELITE
$145

H-53O CA194O
WOOD
$11O

H-531 CA194O
FAUX LEATHER COVERED WOOD
$125

H-62OU CA194O
BAKELITE
$75

HJ-514 CA194O
WOOD
$4O

HJ-612 CA194O
WOOD
$35

HJ-634 CA194O
WOOD
$55

HJ-632U CA194O
WOOD
$5O

HM-8O CA194O
WOOD
$9O

J-51 CA1941
WOOD
$25

J-53 CA1941
WOOD, INGRAHAM CABINET
$85

J-54 CA1941
IVORY PLASKON
$40

J-62 CA1941
WOOD
$40

J-64 CA1941
WOOD
$50

J-70 CA1932
WOOD
$160

J-71 CA1941
WOOD
$80

J-72 CA1933
WOOD
$225

J-80 CA1932
WOOD
$275

J-82 CA1933
Wood
$275

J-83 CA1933
Wood
$165

J-100 CA1933
Wood
$235

J-501 · CA1941
Bakelite
$30

JB-420 CA1941
Bakelite
$30

JB-520 CA1941
Wood
$15

JFM-90 CA1941
Wood
$70

K-40 CA1933
Wood
$145

K-43 CA1933
Wood
$225

K-48 CA1933
2-Tone Wood Radio-Phono
$225

K-52 CA1933
2-Tone Wood
$190

K-53 CA1933
Wood
$125

K-53M CA1933
Wood with Blacl Lacquer Trim
$325

K-60 CA1933
Wood
$225

K-63 CA1933
Wood
$225

K-64 CA1933
Wood
$250

K-80 CA1934
Wood
$250

L-50 CA1933
Repwood
$225

L-53 CA1933
2-TONE WOOD
$160

L-570 CA1942
CATALIN
$500-$2000

L-600 CA1942
IVORY PLASKON
$40

L-613 CA1942
WOOD
$25

L-621 CA1942
BAKELITE
$30

L-630 CA1942
WOOD
$30

L-632 CA1942
WOOD
$25

L-633 CA1942
WOOD, INGRAHAM CABINET
$125

L-640 CA1942
WOOD
$30

L-652 CA1942
Wood
$40

L-740 CA1942
Wood
$45

M-41 CA1933
Wood
$160

M-51 CA1935
Wood
$125

M-61 CA1936
Wood
$250

M-81 CA1935
Wood
$275

'Midget' CA1932
Wood
$150

S-22 'Junior' CA1932
Wood w/Bouquet Embroidered Grille Cloth
$175

S-22X 'Junior' w/Clock CA1932
Wood w/Bouquet Embroidered Grille Cloth
$250

250A 'LITTLE GENERAL' CA1932
PAINTED WOOD
$250

GMR REMOTE CA1932
BRASS PLATED METAL
$175

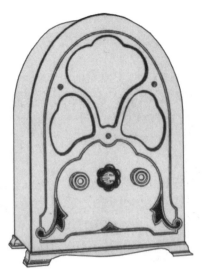

'LITTLE CORPORAL' CA1932
WOOD
$175

'LITTLE GENERAL' CA1932
WOOD
$250

1A5 'BLUE CROWN' CA1948
LEATHERETTE
$25

526 CA1947
WOOD WITH INLAID MARQUETRY
$50

5A5A 'SPY' CA1938
WOOD WITH BAKELITE TRIM
$600+

'A' CA1948
CANVAS BAORD
$25

'ARLINGTON' CA1941
WOOD
$30

'TELETONE' CA1939
BAKELITE
$150

'BABY EMERALD' CA1940
WOOD
$500+

56B CA1947
WOOD
$25

58M CA1947
BAKELITE
$175

68F CA1947
WOOD
$25

CA1934
EXOTIC WOODS
$225

CA1936
WOOD
$250

CA1938
IVORY PLASKON
$75

60 CA1933
WOOD
$225

80 CA1933
WOOD
$190

309L CA1938
WOOD
$150

370C CA1937
WOOD
$450+

762L CA1937
WOOD
$225

782L CA1937
WOOD
$125

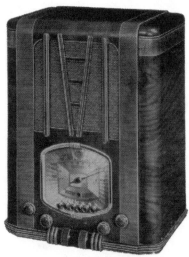

782T CA1937
WOOD
$150

CHALLENGER 1, 2 CA1937
BAKELITE $225, IVORY PLASKON $300
GREEN, RED PLASKON $500+

CHALLENGER 3 CA1937
BAKELITE $225, IVORY PLASKON $300
GREEN, RED PLASKON $500+

CHALLENGER 3C CA1937
WOOD
$135

CHALLENGER 5 CA1937
BAKELITE $225, IVORY PLASKON $300
GREEN, RED PLASKON $500+

'SYNCHRONETTE' CA1933
WOOD WITH MARQUETRY INLAY
$110

Grunow

2A 'REMOTE' CA1933
BURLE WALNUT
$125

410 CA1938
WOOD
$135

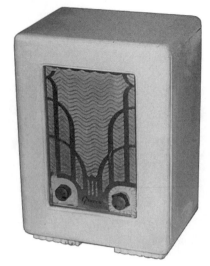

451 CA1935
IVORY LACQUER WOOD WITH CHROME GRILLE
$325

460 CA1934
WOOD
$85

470 CA1937
WOOD
$75

502 CA1934
WOOD
$115

508 CA1938
WOOD
$120

502A CA1937
WOOD
$125

532 CA1935
2-TONE WOOD
$65

544 CA1938
WOOD
$50

566 CA1937
WOOD
$90

576 CA1938
WOOD
$60

580 CA1937
WOOD
$120

592 CA1938
WOOD WITH CHROME GRILLE & TRIM
$175

594 CA1938
IVORY LACQUER WOOD WITH CHROME GRILLE & TRIM
$190

596 CA1938
WOOD
$55

622 CA1938
WOOD WITH BRASS TRIM
$80

GRUNOW 624 CA1938
WOOD WITH BRASS TRIM
$110

650 CA1935
WOOD
$250

654 CA1938
WOOD
$275

660 CA1934
WOOD
$150

670 CA1934
WOOD
$150

680 CA1937
WOOD
$130

700 CA1934
WOOD WITH CHROME GRILLE
$275

750 'WORLD CRUISER' CA1934
WOOD
$275

1101W REMOTE CA1934
WOOD
$325

130 CA1932
WOOD WITH REPWOOD GRILLE
$300

330 CA1932
WOOD
$250

3521 CA1932
WOOD
$250

M5A1 CA1933
WOOD WITH MARQUETRY INLAY
$175

hALSON

5OR CA1936
2-TONE WOOD
$80

62O CA1935
WOOD
$110

'JEFFERSON' CA1934
BLONDE WOOD WITH BLACK LACQUER
$350

T5 CA1937
WOOD
$120

4BT CA1939
WOOD
$30

6 CA1936
WOOD
$115

200 CA19349
WOOD & METAL
$50

220 CA1939
WOOD & METAL
$50

225 CA1938
WOOD
$70

250 CA1938
WOOD
$60

300 CA1939
WOOD
$35

303 CA1939
WOOD
$35

CA1940
WOOD
$40

HOWARD

368 CA1938
WOOD
$60

430 CA1939
WOOD & METAL
$60

468 CA1939
WOOD
$65

575 CA1939
WOOD
$35

901W CA1946
WOOD
$45

'AUTOMATIC' CA1938
WOOD
$85

COMPACT CA1933
WOOD
$115

D15 CA1935
WOOD
$135

'WORLD SEVEN' CA1936
WOOD
$130

47A CA1935
2-TONE WOOD
$65

57A CA1935
WOOD WITH INLAID MARQUETRY
$150

504 CA1936
WOOD
$110

522 CA1936
WOOD
$65

726 CA1936
WOOD
$145

736 CA1936
WOOD
$110

846 CA1936
WOOD
$145

6442 CA1936
WOOD
$125

7767 CA1936
WOOD
$135

25 'PETER PAN' CA1933
WOOD
$500+

25AV 'PETER PAN' CA1933
WOOD
$700+

26SW CA1933
WOOD
$225

38 CA1933
WOOD
$300

60 CA1930
2-TONE WOOD
$275

62 'PEACOCK CA1931
WOOD WITH BLACK LACQUER GRILLE
$500+

68 'SUNRISE' CA1931
WOOD WITH BLACK LACQUER GRILLE
$325

84 'PETER PAN' CA1933
WOOD WITH BLACK LACQUER INSERT GRILLE
$500+

88 'TULIP' CA1933
WOOD WITH BLACK LACQUER GRILLE
$500+

300 CA1940
MARBELIZED PLASTIC
$125

505 'PIN-UP' CA1950
WHITE PLASTIC
$75

814 'TEE-NEE' CA1950
PLASTIC
$125

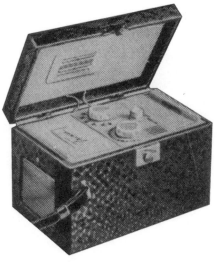

905 'TRIXIE' CA1950
FAUX ALLIGATOR
$50

935 'WAKEMASTER' CA1949
IVORY PLASKON
$40

955 'NUGGET' CA1950
IVORY PLASKON
$60

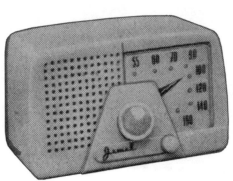

956 CA1950
IVORY PLASKON
$35

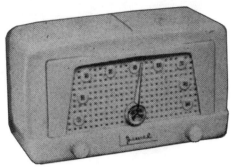

960 CA1950
IVORY PLASKON
$30

R188 CA1940
WOOD WITH ALUMINUM GRILLE
$350+

KADETTE

26 CA1936
2-TONE WOOD
$150

52 CA1936
2-TONE WOOD
$125

53 CA1936
2-TONE WOOD
$135

61 CA1936
2-TONE WOOD
$140

617 CA1937
WOOD
$90

630 CA1938
2-TONE WOOD
$45

634 CA1937
WOOD
$50

635 CA1938
WOOD
$110

735 CA1938
2-TONE WOOD
$50

739 CA1938
WOOD
$55

845 CA1938
WOOD
$55

950 CA1938
WOOD
$70

1019 CA1937
WOOD
$65

1030 CA1938
WOOD
$65

1035 CA1938
WOOD
$70

1129 CA1937
2-TONE WOOD
$75

1140 CA1938
2-TONE WOOD
$75

'AUTIME RADIO ALARM' CA1938
WOOD CLOCK-RADIO
$90

KADETTE

'Duo' CA1934
WOOD
$350

'International Midget' CA1931
WOOD
$250

K-25 'Clockette' CA1937
CATALIN
$1200-8000+

L-25 'Topper' CA1939
IVORY PLASKON
$500+

L-29 'Topper' CA1939
WOOD
$350+

Shortwave Converter CA1935
WOOD
$125

T-61 CA1932
WOOD
$135

55 CA1934
WOOD
$160

500A CA1935
2-TONE WOOD
$110

600A CA1934
WOOD
$150

700A CA1935
WOOD
$90

'GIPSY' CA1934
2-TONE WOOD
$65

9701 H CA1936
WOOD
$130

9703H CA1936
WOOD
$85

9705 CA1936
WOOD
$60

9707H CA1936
WOOD
$135

9709H CA1936
METAL WITH FAUX WOOD GRAIN FINISH
$110

9711H CA1936
WOOD
$85

9715H CA1936
WOOD
$135

9735H CA1936
2-TONE WOOD
$115

9747H CA1936
WOOD
$140

9755H CA1936
2-TONE WOOD WITH MARQUETRY INLAY
$125

9759H CA1936
WOOD
$135

9771H CA1936
WOOD
$90

9773H CA1936
2-TONE WOOD
$125

9780H CA1936
WOOD
$90

9785H CA1936
WOOD
$150

9791 CA1936
WOOD
$155

10800 CA1939
2-TONE WOOD
$140

10801 'PEE WEE' CA1939
IVORY PLASKON
$165

KNIGHT

10803 CA1939
WOOD
$40

10804 CA1939
2-TONE WOOD
$100

10805 CA1939
WOOD
$35

10806 CA1939
WOOD
$35

10807 CA1939
WOOD
$50

10810 CA1939
WOOD
$45

10830 CA1940
IVORY PLASKON
$130

10835 CA1939
WOOD
$40

10840 CA1939
IVORY PLASKON
$175

10845 'Pee Wee' ca1939
Ivory Plaskon
$225

10860 ca1939
Ivory Plaskon
$75

10874 ca1939
Wood
$50

10882 ca1939
Wood
$55

10893 ca1939
Wood
$60

10894 ca1939
Wood
$60

10895 ca1939
Wood
$60

10896 ca1939
Wood
$45

10898 ca1939
Wood
$60

10900 CA1939
WOOD
$55

10905 CA1939
2-TONE WOOD
$110

10906 CA1939
WOOD
$120

10908 CA1939
WOOD
$35

10912 CA1939
2-TONE WOOD
$40

10914 CA1939
WOOD
$45

10920 CA1939
WOOD WITH BLACK LACQUER TRIM
$115

17100 CA1941
WOOD
$25

17101 CA1941
IVORY PAINTED WOOD
$40

17102 CA1941
WOOD
$20

17104 CA1941
IVORY PAINTED BAKELITE
$60

17105 CA1941
WOOD
$25

17106 CA1941
2-TONE WOOD
$30

17110 CA1941
2-TONE WOOD
$35

17126 CA1941
IVORY PAINTED PLASTIC
$45

17127 CA1941
CATALIN
$1000+

17134 'TREASURE CHEST' CA1941
WOOD
$70

17141 CA1941
WOOD
$30

17150 CA1941
BAKELITE
$95

17155 CA1941
2-TONE WOOD
$30

17170 CA1941
WOOD
$35

17171 'FM CONVERTER' CA1941
WOOD
$45

'MIDGET' CA1931
WOOD
$190

3-TUBE CA1939
METAL
$125

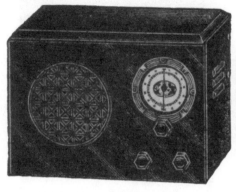

7-C CA1939
METAL
$40

BS-5 CA1939
META;
$40

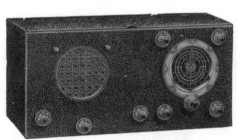

D-38 CA1939
METAL
$45

D-39 CA1939
METAL
$45

A-15 CA1935
WOOD WITH BLACK LACQUER TRIM
$115

A-18 CA1936
WOOD
$95

A-19 CA1935
WOOD
$175

A-22 CA1937
WOOD
$65

A-23 CA1937
WOOD
]$55

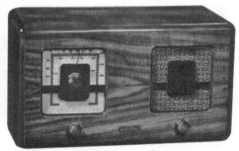

A-41 CA1936
WOOD WITH BLACK METAL GRILLES
$85

LA-70 CA1936
WOOD
$115

AM-26 CA1935
WOOD WITH BLACK LACQUER TRIM
$150

AM-42 CA1935
WOOD
$110

AS-5 CA1936
WOOD MINI-TOMBSTONE
$120

AS-6 CA1936
WOOD
$200

B-21 CA1936
WOOD
$80

B-36 CA1936
WOOD
$80

B-49 CA1939
BAKELITE
$85

B-61 CA1935
WOOD
$160

B-64 CA1935
WOOD
$120

B-81 CA1937
WOOD
$70

BA-4 CA1938
BAKELITE
$50

BA-19 CA1937
WOOD
$45

C-15 CA1939
WOOD
$40

C-16 CA1939
WOOD WITH BLACK LACQUER TRIM
$45

C-20 CA1935
WOOD
$110

C-25 CA1936
WOOD
$110

C-29 CA1939
WOOD
$45

C-35 CA1935
2-TONE WOOD
$95

C-37 CA1939
WOOD
$90

C-39 CA1939
BAKELITE $225, IVORY PLASKON $300
GREEN, RED PLASKON $500+

C-48 CA1937
Wood
$55

C-51 CA1939
Wood
$40

C-55 CA1935
Wood
$110

C-57 CA1938
Wood
$45

C-61 CA1937
Wood
$90

C-84 CA1936
Wood
$75

C-87 CA1939
Wood
$40

C-90 CA1939
Wood
$45

C-91 CA1939
Wood
$50

C-98 CA1939
WOOD
$50

D-2 CA1939
WOOD
$50

D-7 CA1936
WOOD
$300+

D-8 CA1936
WOOD
$40

D-10 CA1937
WOOD
$60

D-11 CA1937
WOOD
$65

D-13 CA1937
WOOD WITH BLACK LACQUER TRIM
$45

D-14 CA1937
WOOD
$90

D-15 CA1937
WOOD
$55

D-16 CA1936
WOOD
$60

D-21 CA1938
WOOD
$55

D-23 CA1938
WOOD
$50

D-24 CA1939
IVORY PLASKON
$225

D-27 CA1937
WOOD
$40

D-28 CA1937
WOOD
$65

D-31 CA1937
WOOD
$65

D-35 CA1938
WOOD
$50

D-41 CA1938
WOOD
$45

D-43 CA1939
IVORY PLASKON
$325

D-47 CA1936
WOOD
$80

D-48 CA1938
WOOD
$45

D-58 CA1939
IVORY PLASKON
$250

DA=13P CA1937
WOOD WITH BLACK LACQUER TRIM
$40

DA-16 CA1937
WOOD
$45

DA-28 CA1937
WOOD
$45

E-191 CA1940
FAUX LEATHER
$90

EB-7 CA1937
WOOD
$60

EB-8 CA1936
WOOD
$75

EB-52 CA1938
WOOD
$50

EM-70 CA1938
IVORY PLASKON
$275

F-52 CA1935
WOOD
$160

F-60 CA1935
2-TONE WOOD
$175

FA-6 CA1937
IVORY PAINTED WOOD
$40

FS-7 CA1938
WOOD
$60

FS-17 CA1938
WOOD
$60

FS-22 CA1938
WOOD
$55

J-11 CA1936
2-TONE WOOD
$75

J-17 CA1936
WOOD
$55

J-32 CA1937
WOOD
$25

J-35 CA1937
WOOD
$30

J-38 CA1936
WOOD WITH MARQUETRY INLAY
$20

J-43 CA1937
WOOD
$55

J-43 CA1937
WOOD
$30

J-79 CA1936
WOOD
$175

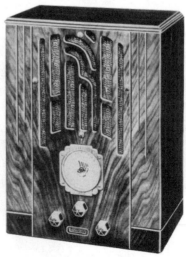

J-85 CA1936
WOOD
$110

JA-7 CA1939
WOOD
$35

JA-35 CA1937
WOOD
$40

JB-3 CA1936
WOOD WITH MARQUETRY INLAY
$75

M-9 CA1936
WOOD
$120

M-31 CA1938
WOOD
$95

M-31 CA1937
WOOD WITH MARQUETRY INLAY
$125

M-43 CA1936
WOOD WITH MARQUETRY INLAY
$125

M-91 CA1938
WOOD
$60

'PACESETTER' CA1934
2-TONE WOOD WITH METAL GRILLE
$325

S-63 CA1936
WOOD
$130

S-107 CA1940
CATALIN
$500+

'THRILLER' CA1932
WOOD
$175

CA1932
WOOD
$225

4626 CA1936
WOOD
$80

ENSIGN CA1933
WOOD
$160

LEUTENANT GOVERNOR CA1934
WOOD
$140

M4616 CA1936
WOOD
$85

Q5636 CA1936
WOOD
$80

185

LYRIC

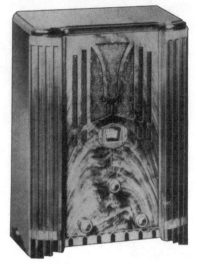

45OA CA1934
WOOD
$250

C-3/M-4 (R.WRIGHT DES) CA1934
BLACK LACQUER WOOD WITH ALUMINUM TRIM
$1000+

C4L CA1934
METAL
$150

C4L1 (RUSSELWRIGHT DES) CA1934
WOOD, BIRDSEYE MAPLE
$450

J3 'JUNIOR' CA1932
WOOD
$325

M4L (RUSSELWRIGHT DES) CA1934
WOOD
$400

MIDGET CA1932
WOOD
$225

P5L M4L (RUSSELWRIGHT DES) CA1934
IVORY PAINTED WOOD WITH CHROME GRILLE
$400

S-6 CA1931
WOOD
$225

S6 CA1932
WOOD
$250

S-7 CA1931
WOOD
$265

SA5L (RUSSELWRIGHT DES) CA1934
WOOD
$400

SU5L (RUSSELWRIGHT DES) CA1934
WOOD WITH IVORY LACQUER TRIM
$450+

SW88 (RUSSELWRIGHT DES) CA1934
22-TONE WOOD
$750+

U55 CA1933
2-TONE WOOD
$150

U5L1 CA1934
WOOD
$175

1A59 CA1939
2-TONE WOOD
$65

5ADA CA1940
WOOD
$75

5BDA CA1940
2-TONE WOOD
$50

5TO CA1940
IVORY PLASKON WITH MIRROR DIAL CLOCK-RADIO
$350

42A 'SPECIAL' CA1934
WOOD WITH MARQUETRY INLAY
$100

50 CA1937
WOOD
$135

56 CA1938
WOOD WITH BLACK LACQUER TRIM
$150

57 CA1938
WOOD WITH BLACK LACQUER TRIM
$110

60 CA1937
2-TONE WOOD
$175

61 CA1938
Wood
$115

65 CA1937
Wood
$250

67 CA1938
Wood with Black Lacquer Trim
$175

68 CA1938
Wood
$70

75 CA1938
Wood
$275

75A CA1934
Wood
$225

85 CA1938
Wood
$350

86 CA1937
Wood
$190

194 CA1934
Wood with Chrome Bezel
$250

CA1934
Wood
$250

196 CA1934
Wood
$250

259EB CA1939
2-Tone Wood
$90

291 CA1933
Wood
$325

311 CA1933
Wood
$250

331 'Gothic' CA1933
Wood
$375

371 CA1932
Wood
$175

373 CA1932
Wood
$350

381 'Treasure Chest' CA1933
Wood with Repwood Trim
$165

422 CA1934
WOOD
$120

428 CA1934
WOOD
$110

521 CA1934
WOOD
$135

523 CA1934
WOOD WITH BLACK LACQUER TRIM
$150

631 CA1934
WOOD WITH MARQUETRY INLAY
$115

669 CA1934
2-TONE WOOD WITH CHROME ESCUTCHEONS
$500+

1583W 'COIN-OP' CA1950
IVORY PAINTED BAKELITE
$125

9160 CA1955
IVORY PAINTED BAKELITE
$50

A457 'GOTHIC' CA1934
2-TONE WOOD
$225

CA1933
WOOD
$115

CA1933
WOOD WITH INLAY
$135

CA1935
2-TONE WOOD WITH CHROME GRILLE INSERT
$225

M204 'DELUXE' CA1934
2-TONE WOOD
$160

'MAYFAIR' CA1934
WOOD WITH BLACK LACQUER & ALUMINUM
$325

EXTENSION SPEAKER CA1933
WOOD WITH BLACK LACQUER & ALUMINUM
$175

T-081A CA1941
WOOD
$60

CA1941
WOOD
$80

5C5-DW9 'Trail Blazer' ca1946
Wood
$40

5C5-P12 'Trail Blazer' ca1946
Ivory Painted Bakelite
$60

CW500 ca1946
Ivory Painted Bakelite
$55

'Trail Blazer Table' ca1947
Wood with Marquetry Inlay
$45

9-1053 CA1942
WOOD
$30

'BREWSTER' 6D CA1948
IVORY PAINTED BAKELITE
$45

'FM' CA1941
WOOD
$45

'ROBOT CONTROL' CA1939
BAKELITE
$225

'ROBOT CONTROL' CA1938
WOOD
$150

98 CA1941
WOOD
$150

BB-7 CA1938
WOOD
$150

CC-8 CA1938
WOOD
$225

CTC 'ALL WAVE CONVERTER' CA1932
WOOD
$125

D-7 CA1937
WOOD
$250

G-10 CA1934
WOOD
$450+

G-10 CA1935
WOOD
$450+

G-11 CA1936
WOOD
$450+

GG-9 CA1938
WOOD
$350+

H-5 CA1936
WOOD
$175

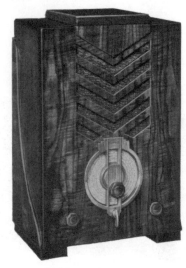

HH-7 CA1936
WOOD
$425+

II-10 CA1938
WOOD
$400+

K-11 CA1937
WOOD
$400+

LL-12 CA1938
WOOD
$600+

X-5 CA1933
2-TONE WOOD
$300

5C 'Radio-Larm' ca1951
Painted Bakelite Clock-Radio
$65

5H ca1951
Painted Bakelite
$50

5J 'Jewel Box' ca1951
Bakelite with Green Marble Plastic Grille
$115

5L 'Music Box' ca1951
Painted Bakelite
$50

5R ca1951
Painted Bakelite
$60

5T1 ca1937
Wood
$45

5T11G ca1959
Colored Plastic
$45

5T2 ca1937
Wood
$250+

5X ca1951
Bakelite with Metal Stand & Red Tenite Trim
$110

6L1 'Towne & Country' ca1951
Colored Plastic
$55

6T ca1937
Wood
$60

6X ca1937
2-Tone Wood
$50

40W ca1940
Wood
$35

41B-11 ca1942
Bakelite
$25

41B-12 ca1940
Wood
$25

41E ca1940
Wood
$25

50XC 'Circle Grille' ca1940
Catalin
$1500+

51A ca1940
Bakelite
$150

51X-11 CA1942
BAKELITE
$45

51X-16 'S-Grille' CA1942
CATALIN
$3000+

52C 'Vertical Grille' CA1940
CATALIN
$1500+

52H CA1952
PAINTED BAKELITE
$55

52T CA1937
WOOD
$70

53A CA1940
WOOD
$175

53LC CA1953
PLASTIC
$125

56C-5 CA1956
COLORED UREA
$110

56T CA1937
WOOD
$55

57CS c1957
Colored Urea
$110

59H-11 ca1950
Bakelite
$55

59L-12 ca1950
Bakelite
$60

59R11 ca1950
Painted Bakelite
$55

59T-4 ca1939
Wood
$75

59T-5 ca1939
Wood
$55

59X-11 ca1950
Bakelite
$60

61A ca1940
Bakelite
$160

61C ca1940
Wood
$65

61T-21 CA1942
WOOD
$30

6T-22 CA1942
WOOD
$35

61T-23 CA1942
WOOD
$35

61X-12 CA1942
IVORY PLASKON
$45

61X-13 CA1942
WOOD
$30

61X-1 CA1942
WOOD
$30

63E CA1940
WOOD
$70

63L CA1953
COLORED PLASTIC
$45

65X-13 CA1946
2-TONE WOOD
$40

75T-31 CA1946
WOOD
$45

496BT-1 CA1946
WOOD
$30

A1R-23 CA1959
COLORED PLASTIC
$50

A4G-23 'CUSTOM SIX' CA1956
COLORED PLASTIC & PAINTED METAL
$40

S-10 'LAZY BOY CONTROL' CA1934
2-TONE WOOD WITH BLACK LACQUER TRIM
$175

S-10 'LAZY BOY SPEAKER' CA1934
2-TONE WOOD WITH BLACK LACQUER TRIM
$400+

S-10 'LAZY BOY SPEAKER' CA1934
2-TONE WOOD WITH BLACK LACQUER TRIM
$400+

1A1 CA1938
IVORY PLASKON
$110

2S1 CA1938
WOOD
$50

2S4 CA1938
WOOD
$40

2S7 CA1938
WOOD
$80

3A2 CA1938
IVORY PLASKON $135
BEETLE PLASTIC $300+

3H5 CA1938
WOOD
$45

3S3 CA1938
WOOD
$60

4N2 CA1938
WOOD
$65

14H3 CA1938
WOOD
$115

15M2 CA1938
WOOD (10 TUBES)
$500+

21M1 CA1938
WOOD
$50

44 CA1935
2-TONE WOOD WITH INLAID MARQUETRY
$80

200 CA1938
WOOD
$70

P12EN1 CA1938
WOOD
$65

5AC CA1939
WOOD
$65

5D 'KOMPAK' CA1939
IVORY PLASKON
$70

5E 'DELUXE' CA1939
WOOD
$110

5N CA1939
WOOD
$70

5N 'DELUXE' CA1939
WOOD
$150

35N 'DELUXE' CA1939
PAINTED WOOD
$175

45A CA1935
WOOD
$225

46E 'DELUXE' CA1939
WOOD
$150

46G 'DELUXE' CA1939
WOOD
$160

48D 'Deluxe' ca1939
Wood
$150

48E 'Deluxe' ca1939
Wood
$160

872 'FM Converter' ca1948
Wood
$50

90 CA1933
WOOD
$150

308 CA1937
WOOD
$125

312 CA1937
WOOD
$135

MIDGET CA1931
WOOD
$225

PHILCO

6 CA1939
WOOD
$45

7T CA1938
WOOD
$125

9 CA1938
WOOD
$90

1OT 'BULLET' CA1938
WOOD
$125

12 CA1938
WOOD
$45

12C CA1938
WOOD
$55

15T CA1938
WOOD
$50

16B CA1933
WOOD
$175

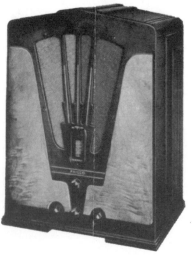

16B CA1933
WOOD
$250

18B CA1933
WOOD
$185

18B CA1934
WOOD
$225

19B CA1933
WOOD
$185

19T CA1939
WOOD
$45

20 CA1930
WOOD
$175

20 'DELUXE' CA1930
WOOD
$225

21B CA1930
WOOD
$275

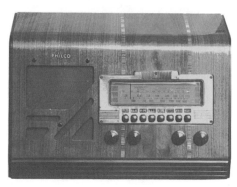

25 CA1939
WOOD
$65

32 CA1934
WOOD
$200

PHILCO

33B CA1938
Wood
$55

34B CA1938
Wood
$75

38B CA1934
2-Tone Wood
$150

38T 'Bullet' CA1938
Wood
$125

39T CA1938
Wood
$90

43B CA1932
Wood
$175

44B CA1934
Wood
$165

50B CA1932
Wood
$160

54C CA1934
Wood
$110

54C CA1935
WOOD
$110

54C CA1935
2-TONE WOOD
$115

54S CA1934
WOOD WITH FAUX FINISH
$150

57C CA1933
WOOD
$90

50B CA1938
WOOD
$140

60 CA1933
WOOD
$160

60B CA1935
WOOD
$140

60B CA1935
2-TONE WOOD
$175

61B CA1937
WOOD
$125

62T CA1938
WOOD WITH FAUX FINISH
$80

66 CA1934
2-TONE WOOD
$175

66B CA1935
WOOD
$125

70 CA1932
WOOD
$325

70 CA1939
WOOD
$75

71B CA1932
WOOD
$185

84B CA1935
2-TONE WOOD WITH BLACK LACQUER TRIM
$125

84B CA1936
2-TONE WOOD
$75

84B CA1937
WOOD
$75

89 CA1932
2-TONE WOOD
$170

89B CA1935
2-TONE WOOD
$175

89B CA1936
WOOD
$75

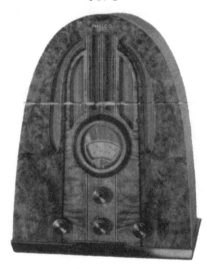

89B CA1937
WOOD
$80

90 CA1931
WOOD
$450

90 CA1940
BAKELITE
$45

90 CA1941
BAKELITE
$40

91B CA1932
WOOD
$200

93 CA1938
WOOD
$65

93 CA1938
WOOD
$55

95 CA1940
WOOD
$45

100 CA1940
WOOD
$50

105 CA1940
WOOD
$75

110 CA1940
WOOD
$80

115 CA1940
WOOD
$65

116B CA1936
WOOD
$70

116B CA1937
WOOD
$125

118 CA1934
WOOD
$225

120 CA1949
WOOD
$30

124 CA1940
WOOD
$40

125 CA1940
WOOD
$30

135 CA1940
WOOD
$45

145 CA1940
WOOD
$45

150 CA1940
WOOD
$55

155 CA1940
WOOD
$55

200 CA1946
BAKELITE
$30

201 CA1946
WOOD
$25

PHILCO

203 CA1946
WOOD WITH IVORY PAINT TRIM
$60

220 CA1941
2-TONE WOOD
$45

221 CA1941
WOOD WITH MARBLED TENITE GRILLE
$75

225 CA1941
WOOD
$30

230 CA1941
BAKELITE
$35

231 'BULLET' CA1941
WOOD
$115

235 CA1941
WOOD WITH BLACK LACQUER TRIM
$30

240 CA1941
WOOD WITH BLACK LACQUER TRIM
$35

245 CA1941
WOOD WITH BLACK LACQUER TRIM
$40

250 CA1941
WOOD
$55

250 CA1946
BAKELITE
$35

251 CA1946
WOOD
$25

255 CA1941
WOOD
$55

321 CA1942
WOOD WITH IVORY PAINT TRIM
$40

322 CA1942
WOOD
$45

323 CA1942
BAKELITE
$35

327 CA1942
WOOD WITH BLACK LACQUER TRIM
$35

340 CA1942
WOOD
$35

345 CA1942
Wood
$35

350 CA1942
Wood
$135

355 CA1942
Wood
$45

420 CA1946
Painted Bakelite
$50

421 CA1946
Wood
$25

422 CA1946
Wood with Ivory Paint Trim
$30

427 CA1946
Wood
$30

431 CA1946
Wood
$35

433 CA1946
Wood
$35

452 CA1946
WOOD
$50

454 CA1946
WOOD
$25

489 CA1940
WOOD
$30

520 CA1950
BAKELITE
$20

522 CA1950
BAKELITE
$25

524 CA1950
WOOD
$20

532 CA1951
PAINTED BAKELITE
$60

600C CA1937
WOOD WITH FAUX FINISH
$65

602C CA1936
WOOD
$65

604C CA1037
WOOD WITH BAKELITE TRIM
$250+

610 CA1936
WOOD
$75

610 'BIG BULLET' CA1937
WOOD
$145

610B CA1937
WOOD
$80

610B CA1937
WOOD WITH FAUX FINISH
$75

610T 'BIG BULLET' CA1936
WOOD
$145

620B CA1937
WOOD
$125

620B CA1937
WOOD WITH FAUX FINISH
$70

625B CA1936
WOOD
$70

63OT CA1937
WOOD
$80

64OB CA1937
WOOD WITH FAUX FINISH
$70

643B CA1937
WOOD
$150

655B CA1936
WOOD
$90

66OB CA1937
WOOD
$110

67OB CA1937
WOOD
$65

714 CA1954
PAINTED BAKELITE CLOCK-RADIO
$35

901 'SECRETARY' CA1939
BAKELITE
$225

902 CA1949
BAKELITE WITH PLASKON GRILLE
$45

921 CA1950
PAINTED BAKELITE
$35

922 CA1950
BAKELITE
$30

925 CA1950
WOOD
$20

926 CA1950
BAKELITE
$25

6015 CA1939
WOOD
$90

C579 CA1955
PLASTIC
$30

J775-124 'PREDICTA' CA1958
PLASTIC
$150

PT-4 CA1942
PAINTED BAKELITE
$30

PT-7 CA1942
WOOD
$30

PT-10 CA1942
BAKELITE
$30

PT-25 CA1942
BAKELITE
$50

PT-38 CA1939
WOOD
$25

PT-48 CA1939
IVORY PLASKON
$110

PT-50 CA1939
WOOD
$30

PT-66 CA1939
WOOD WITH IVORY PAINTED TRIM
$40

PT-69 CA1939
WOOD CLOCK-RADIO
$115

PT-91 CA1942
BAKELITE
$30

PT-93 CA1942
WOOD
$60

PHILCO

PT-94 CA1942
WOOD WITH TENITE GRILLE
$75

PT-95 CA1942
WOOD WITH TENITE GRILLE
$75

TH-14 CA1939
WOOD
$35

TH-16 CA1939
BAKELITE
$40

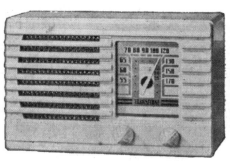

TP-4 CA1939
IVORY PAINTED BAKELITE
$40

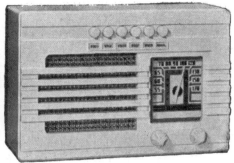

TP-5 CA1939
IVORY PAINTED BAKELITE
$50

TP-11 CA1939
BAKELITE WITH TENITE GRILLE
$350+

TP-12 CA1939
WOOD
$50

TP-15 CA1939
WOOD WITH BLACK LACQUER GRILLE
$35

43 CA1934
WOOD
$275

53 CA1935
WOOD
$225

63 CA1935
WOOD
$250

304 CA1937
WOOD
$325

601 'PILOTUNER FM' CA1948
WOOD
$20

'ALLWAVE' CA1932
WOOD
$450+

B-2 CA1934
WOOD
$300

'BROADCAST' CA1933
WOOD
$275

'DRAGON' 18 CA1933
WOOD
$300

'DRAGON' CA1934
WOOD
$325

'GOTHIC' CA1931
WOOD
$275

MIDGET CA1931
2-TONE WOOD
$350

MIDGET CA1931
WOOD
$350

'MODERNISTIC' CA1932
BLONDE WOOD WITH BLACK LACQUER TRIM
$400+

SW CONVERTER CA1932
WOOD
$175

TG528 CA1938
WOOD
$125

TRF MIDGET CA1932
WOOD
$250

X203AB CA1937
IVORY PLASKON WITH GOLD TRIM
$1000+

582M CA1935
2-TONE WOOD
$70

584A CA1935
2-TONE WOOD
$90

584B CA1935
2-TONE WOOD
$80

586D CA1935
2-TONE WOOD
$175

586DC CA1935
2-TONE WOOD WITH ALUMINIM GRILLE
$275

588M CA1935
WOOD
$65

592A CA1935
WOOD WITH MARQUETRY INLAY
$135

594M CA1935
2-TONE WOOD
$135

596DL CA1935
2-TONE WOOD
$115

598DC CA1935
WOOD WITH SILVER TRIM & ALUMINUM GRILLE
$225

5100A CA1935
WOOD
$110

5100B CA1935
WOOD WITH MARQUETRY IN LAY
$125

5100C CA1935
2-TONE WOOD
$135

5122AW5 CA1935
WOOD
$110

5514A CA1935
2-TONE WOOD
$115

5514B CA1935
2-TONE WOOD
$145

913 CA1936
Wood
$115

924 CA1936
Wood
$140

925 CA1936
Wood
$110

927 CA1936
Wood
$125

928 CA1936
Wood
$140

929 CA1936
Wood
$135

931 CA1936
Wood
$125

933 CA1936
Wood
$90

936 CA1936
Wood
$115

938 CA9136
Wood
$135

940 CA1936
Wood
$110

943 CA1936
Wood
$125

947 CA1936
Wood
$250

950 CA1936
Wood
$115

953 CA1936
Wood
$115

956 CA1936
Wood
$70

965 CA1936
Wood
$225

966 CA1936
Wood
$110

986 CA1936
BAKELITE $1000+
PLASKON $1500+

6760 CA1937
WOOD
$45

6763 CA1937
WOOD
$85

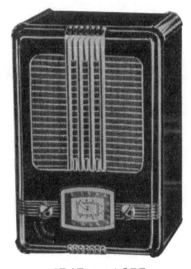

6767 CA1937
BKELITE WITH CHROME TRIM
$225

1-R-81 CA1952
Bakelite
$30

1-X-51 'Blaine' CA1953
Colored Plastic
$35

1-X-591 'Gladwyn' CA1952
Bakelite with Gold Grille
$35

2-C-511 CA1953
Colored Plastic Clock-Radio
$35

2-C-251 CA1952
Colored Plastic Clock-Radio
$20

2-R-51 CA1952
2-Tone Colored Plastic
$25

2-X-62 CA1952
Bakelite
$25

2-X-621 CA1952
Bakelite
$25

2-XF-91 'Forbes' CA1952
Colored Plastic Clock-Radio
$25

3-X-521 CA1954
PLASTIC
$15

4-C-621 'PROMPTER' CA1952
COLORED PLASTIC CLOCK-RADIO
$25

4-T CA1937
2-TONE WOOD
$130

4-X-551 'CREIGHTON' CA1954
COLORED PLASTIC
$110

4-X-641 'DRISCLL' CA1954
PLASTIC
$20

5-BT CA1937
2-TONE WOOD
$80

5-C-581 'DEBONAIRE' CA1954
PLASTIC CLOCK-RADIO
$45

5-X-4 CA1937
IVORY & BLACK PAINTED WOOD
$75

5-X-560 'GREENWICH' CA1954
COLORED PLASTIC W/REVERSE-PAINTED GOLD BEZEL
$50

6-BT CA1937
WOOD
$110

6-BT-6 CA1937
WOOD
$110

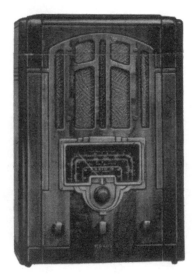

6-T-2 CA1937
WOOD
$110

6-XF-9 'LINDSAY' CA1955
BLACK PLASTIC
$25

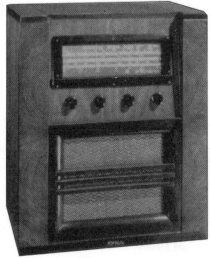

8-QB CA1939
WOOD
$60

8-T CA1938
WOOD
$135

9-BX-56 CA1949
COLORED PLASTIC
$40

9-T CA1936
WOOD
$150

9-X-571 'BULLHORN' CA1950
FAUX WOOD GRAINED PLASTIC
$50

9-XL-1F CA1951
COLORED PLASTIC W/CIGARETTE LIGHTER ON TOP
$45

1O-T CA1937
WOOD
$175

15-X CA1941
BAKELITE
$20

16-X-4 CA194O
WOOD
$30

24-BT-2 CA194O
WOOD
$40

25-BT-3 CA194O
WOOD
$45

25-X-4 CA194O
WOOD
$30

28-X-5 CA194O
WOOD
$35

4O-X-3O 'LITTLE NIPPER' CA194O
BAKELITE
$70

40-X-51 'LITTLE NIPPER' CA1940
WOOD
$75

40-X-52 'LITTLE NIPPER' CA1940
IVORY PAINTED WOOD
$75

45-X-1 'LITTLE NIPPER' CA1940
BAKELITE
$70

45-X-11 CA1940
BAKELITE
$40

45-X-13 CA1940
WOOD
$35

46-X-3 CA1940
WOOD
$35

46-X-11 CA1940
BAKELITE
$40

46-X-13 CA1940
WOOD
$40

54-B-5 CA1947
GOLD METAL WITH CATALIN FACE PLATE
$275

65-X-1 CA1948
BAKELITE
$25

66-X-2 CA1948
PAINTED BAKELITE
$35

66-X-3 CA1947
WOOD
$30

66-X-8 CA1947
CATALIN
$400

66-X-11 CA1948
BAKELITE
$30

66-X-14 CA1948
WOOD
$35

68-R-1 CA1948
BAKELITE
$25

75-X-1 CA1948
BAKELITE WITH GOLD PAINTED FACE
$30

84-BT CA1938
WOOD
$90

85-BT CA1938
WOOD
$80

85-BT-6 CA1938
WOOD
$45

85-T CA1938
WOOD
$40

85-T-1 CA1938
WOOD
$45

85-T-2 CA1938
WOOD
$60

86-BT CA1938
WOOD
$60

86-T CA1938
WOOD
$85

86-T-1 CA1938
WOOD
$50

86-T-2 CA1938
WOOD
$85

86-X CA1938
WOOD
$50

87-T-1 CA1938
WOOD
$40

87-T-2 CA1938
WOOD
$55

94-BT CA1939
WOOD
$60

94-BT-1 CA1939
WOOD WITH BLACK LACQUER TRIM
$90

94-BT-2 CA1939
WOOD
$45

96-T CA1939
WOOD WITH BLACK LACQUER TRIM
$55

96-T-2 CA1939
WOOD WITH BLACK LACQUER TRIM
$65

96-T-3 CA1939
WOOD
$40

96-T-4 CA1939
Wood with Chrome Grille & Trim
$80

96-T-7 CA1939
Wood with Chrome Grille & Trim
$90

97-T-2 CA1939
Wood
$50

98-T CA1939
Wood
$45

99-T CA1939
Wood with Black Lacquer Trim
$75

102 CA1933
Silver & Black Painted Metal
$125

103 CA1936
Wood
$100

114 CA1933
2-Tone Wood
$125

115 CA1933
2-Tone Wood
$250

125 CA1936
WOOD
$155

128 CA1933
WOOD
$350

142B CA1938
2-TONE WOOD
$225

300 CA1934
WOOD RADIO-PHONO
$250

301 CA1933
WOOD RADIO-PHONO
$225

501 'RADIOLA' CA1942
IVORY PLASKON
$115

510 'RADIOLA' CA1942
IVORY PAINTED BAKELITE
$40

513 'RADIOLA' CA1942
WOOD
$45

515 'RADIOLA' CA1942
WOOD
$30

516 'Radiola' ca1942
Bakelite
$25

517 'Radiola' ca1942
Wood
$25

527 'Radiola' ca1942
Wood
$30

81OT ca1938
Wood with Black Lacquer Trim
$85

B-50 ca1942
Wood
$30

B-52 ca1942
Wood
$35

BC-7-9 ca1936
Wood
$115

BT-6-10 ca1936
Wood
$150

BT-6-3 ca1936
Wood
$120

BT-6-5 CA1936
Wood
$145

'Nursery' CA1933
Ivory Painted Wood with Color Decals
$750+

PX-600 'Globetrotter' CA1952
Bakelite
$40

R-4 CA1932
Wood
$225

R-7 'Radiolette' CA1932
Wood
$150

R-22-W CA1933
Wood
$150

R-28 CA1933
Wood
$115

R-28-B CA1933
Wood
$140

R-31-B 'Gothic' CA1933
Wood
$300

R-37 CA1933
WOOD
$250

SW CONVERTER CA1932
WOOD
$170

T-4-8 CA1936
WOOD
$110

T-4-9 CA1936
2-TONE WOOD
$55

T-4-10 CA1936
2-TONE WOOD
$80

T-5-2 CA1936
WOOD
$115

T-6-1 CA1936
WOOD
$155

T-6-9 CA1936
WOOD
$120

T-7-5 CA1936
WOOD
$135

T-8-14 CA1936
WOOD
$170

T-10-1 CA1936
WOOD
$250

T-55 CA1940
WOOD
$35

T-60 CA1940
WOOD
$35

T-62 CA1940
WOOD
$35

T-63 CA1940
WOOD
$35

T-80 CA1940
WOOD
$50

'BOOKSET' CA1948
BAKELITE
$1000+

CA1938
PALE PINK PLASKON
$350+

10 CA1932
WOOD
$250

15 CA1932
WOOD
$250

26 CA1935
BAKELITE WITH IVORY PLASKON TRIM
$400+

30 CA1934
WOOD
$165

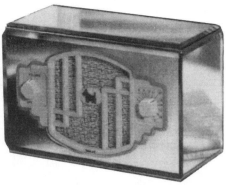

40 'MIRROR CASE' CA1936
MIRROR CABINET WITH IVORY PLASKON GRILLLE
$1500+

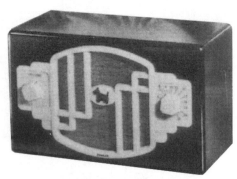

40 CA1936
BAKELITE WITH IVORY PLASKON TRIM
$400+

43 'ESQUIRE' CA1936
WOOD
$225

45 CA1938
WOOD
$75

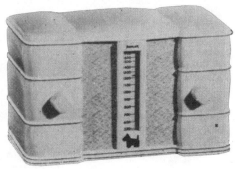

46 CA1937
IVORY PLASKON
$350

51 'Skipper' ca1936
Bakelite $110, Ivory Plaskon $200
Beetle Plastic $325

54 ca1939
Ivory Plaskon
$125

61 ca1939
Wood
$110

88 ca1936
Wood
$425

5300 ca1947
Ivory Plaskon & Bakelite Radio-Phono
$225

D68 ca1939
Wood
$125

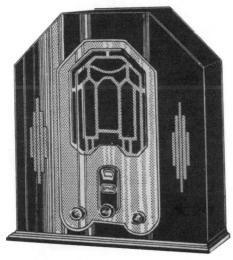

Midget ca1931
Wood
$200

Midget ca1931
Wood
$200

Minuette ca1932
Wood
$275

31B74 CA1936
WOOD
$95

72AT CA1938
WOOD
$55

1O8 CA1931
WOOD
$225

1O8A CA1932
WOOD
$185

6 CA1932
WOOD
$14O

116A CA1932
WOOD
$175

118BCT CA1939
WOOD
$75

125AATE CA1939
WOOD
$5O

137UT CA1939
BAKELITE
$125

177U CA1940
CATALIN
$1000+

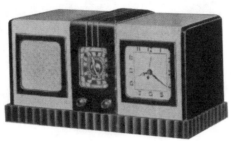

194UTC CA1940
BLONDE WOOD WITH BLACK LACQUER TRIM
$110

194UTI CA1940
PAINTED BAKELITE
$50

195ULTA CA1940
CATALIN
$1200 - $3500

195ULTO CA1940
BEETLE PLASTIC
$325

195ULTWD CA1940
WOOD
$45

198ALT CA1940
WOOD
$65

203UA22 CA1940
WOOD
$110

203ULT CA1940
WOOD
$110

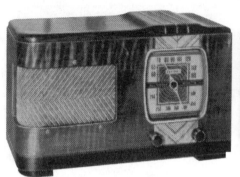

204AT CA1940
WOOD
$65

212 CA1941
BAKELITE
$35

212T CA1941
WOOD
$45

214T CA1941
WOOD
$40

218 CA1941
BAKELITE
$35

218T CA1941
WOOD
$40

220T CA1941
WOOD
$30

221T CA1941
WOOD
$35

226 CA1941
BAKLEITE
$50

236TE CA1941
WOOD
$30

241T CA1941
WOOD
$40

314I CA1947
BAKELITE
$45

329I CA1948
PAINTED BAKELITE
$60

331W CA1949
BAKELITE
$45

332I CA1949
BAKELITE

623 CA1934
WOOD
$165

5721 CA1935
WOOD
$225

B41545 CA1934
WOOD WITH BLACK LACQUER TRIM
$110

37 CA1931
WOOD WITH MARQUETRY INLAY
$155

47A CA1935
2-TONE WOOD
$110

50 CA1935
2-TONE WOOD
$125

57A CA1935
WOOD
$150

68A CA1935
WOOD WITH CHROME TRIM
$140

401H CA1935
2-TONE WOOD
$230

626 CA1935
2-TONE WOOD
$175

736 CA1935
WOOD
$150

846 CA1935
WOOD
$160

1 CA1950
METAL (BY ARVIN)
$50

2 CA1959
PLASTIC
$40

5 CA1959
PLASTIC
$10

6 CA1951
BAKELITE
$35

7 CA1959
PLASTIC
$15

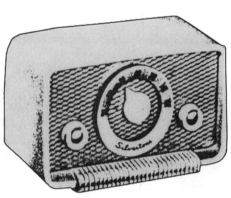

13 CA1951
PLASKON
$40

15 CA1950
BAKELITE
$25

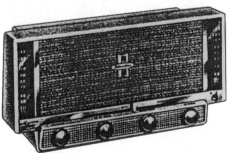

25 CA1959
PLASTIC
$10

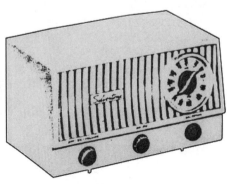

27 CA1952
BAKELITE
$15

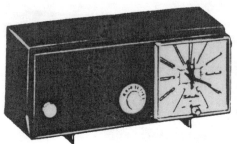

31 CA1959

PLASTIC CLOCK-RADIO
$15

34 CA1959

UREA CLOCK-RADIO
$25

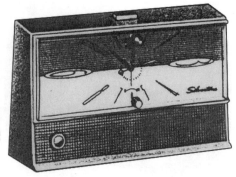

36 CA1959

PLASTIC CLOCK-RADIO
$15

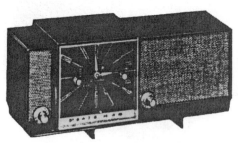

39 CA1959

PLASTIC CLOCK-RADIO
$20

43 CA1959

PLASTIC CLOCK-RADIO
$10

47 CA1959

PLASTIC CLOCK-RADIO (TRANSISTOR)
$25

47 CA1936 (MISSION BELL)

WOOD WITH BLACK LACQUER DETAIL
$350

250 CA1932

WOOD
$175

1002 CA1959

PLASTIC
$10

1007 CA1959
PLASTIC
$10

1013 CA1959
PLASTIC
$10

1022 CA1959
UREA
$35

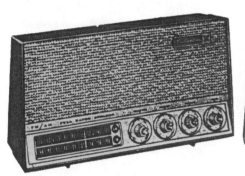

1023 CA1959
PLASTIC
$10

11034 CA1959
PLASTIC CLOCK-RADIO
$15

1036 CA1959
PLASTIC
$20

1039 CA1959
PLASTIC CLOCK-RADIO
$10

1152 CA1932
WOOD WITH REPWOOD TRIM
$250

1290 CA1932
WOOD
$225

1371 CA1932
WOOD
$160

1471 CA1932
WOOD
$200

1481 CA1933
WOOD
$200

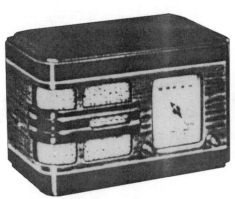

1521 CA1940
WOOD
$40

1561 CA1940
WOOD
$45

1571 CA1933
WOOD
$180

1585 CA1933
WOOD
$225

1589 CA1933
WOOD
$225

1591 CA1933
WOOD
$180

1621 CA1933
WOOD
$150

1660 CA1933
WOOD
$175

1703 CA1934
WOOD
$60

1705 CA1934
WOOD
$75

1706 CA1934
WOOD
$190

1708 'WORLD'S FAIR' CA1934
WOOD WITH SILVER METAL INLAY
$450

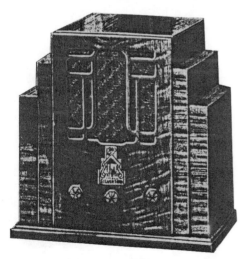

1711 'WORLD'S FAIR' CA1934
WOOD
$375

1712 CA1934
WOOD
$190

1724 CA1934
WOOD
$175

1800 CA1934
WOOD
$170

1801 CA1934
WOOD
$115

1802 CA1935
WOOD
$90

1803 CA1934
WOOD
$95

1804 CA1935
WOOD
$125

1805 CA1935
WOOD
$135

1806 CA1934
2-TONE WOOD
$120

1808 'WORLD'S FAIR' CA1933
WOOD
$325

1810 CA1935
WOOD
$85

1850 'WORLD'S FAIR' CA1934
WOOD
$325

1852 'WORLD'S FAIR' CA1934
WOOD
$325

1855 CA1934
WOOD
$85

1836 CA1935
WOOD
$110

1903 CA1936
WOOD
$85

1923 CA1936
WOOD
$160

1932 CA1936
WOOD
$115

1947 CA1936
WOOD
$140

1954 CA1935
WOOD
$90

2004 CA1954
PLASTIC
$15

2013 CA1954
PLASTIC CLOCK-RADIO
$15

2015 CA1954
PLASTIC
$25

2029 CA1954
WOOD
$10

2611 CA1940
WOOD
$40

2651 CA1940
WOOD
$45

2961 CA1940
WOOD
$55

3002 CA1954
PLASTIC
$45

3005 CA1954
PLASTIC
$20

3007 CA1954
PLASTIC CLOCK-RADIO
$25

3026 CA1954
PLASTIC CLOCK-RADIO
$15

3451 CA1940
WOOD WITH BLACK LACQUER TRIM
$225

3541 CA1939
PLASTIC
$110

4017 CA1954
WOOD
$10

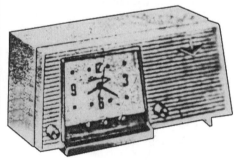

4026 CA1954
PLASTIC CLOCK-RADIO
$15

4200 CA1954
PLASTIC
$15

4205 CA1954
PLASTIC
$15

4418 CA1937
WOOD
$90

Silvertone

4422 CA1937
Wood
$150

4428 CA1937
Wood
$70

4434 CA1937
Wood
$140

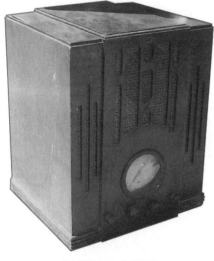

4437 CA1937
Wood
$110

4447 CA1937
Wood
$110

4463 CA1937
Wood
$60

4467 CA1937
Wood
$90

4465 CA1937
Wood
$160

4472 CA1937
Wood
$90

4520 CA1936
2-TONE WOOD
$90

4521 CA1936
2-TONE WOOD
$110

4524 CA1936
WOOD
$90

4526 CA1936
WOOD
$120

4528 CA1936
WOOD
$60

4529 CA1936
WOOD
$125

4532 CA1936
WOOD
$50

4563 CA1936
WOOD
$50

4564 CA1936
2-TONE WOOD
$95

4565 CA1936
WOOD
$125

4566 CA1937
WOOD
$110

4612 CA1939
WOOD
$40

4614 CA1938
WOOD
$60

4620 CA1937
2-TONE WOOD
$60

4622 CA1938
WOOD
$90

4624 CA1938
WOOD
$100

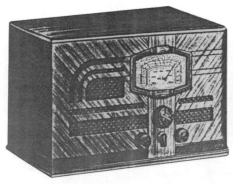

4626 CA1937
WOOD
$65

4630 CA1937
WOOD
$45

4632 CA1938
WOOD
$85

4634 CA1937
WOOD
$40

4636 CA1937
WOOD
$110

4638 CA1938
WOOD
$90

4641 CA1938
WOOD
$35

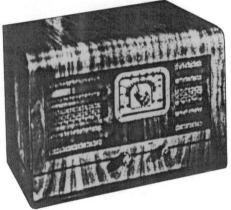

4644 CA1938
WOOD
$40

4660 CA1937
WOOD
$35

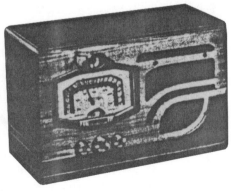

4662 CA1938
WOOD
$40

4663 CA1937
WOOD
$225

4664 CA1937
WOOD
$45

4665 CA1937
WOOD
$155

4666 CA1937
WOOD
$375

4675 CA1938
WOOD
$150

4683 CA1938
WOOD
$45

6003 CA1939
WOOD
$175

6009 CA1946
BAKELITE
$25

6016 CA1947
BAKELITE
$25

6020 CA1955
PLASTIC CLOCK-RADIO
$20

6023 CA1938
WOOD
$40

6042 CA1938
WOOD
$40

6044 CA1938
WOOD
$45

6046 CA1938
WOOD
$45

6048 CA1938
WOOD
$45

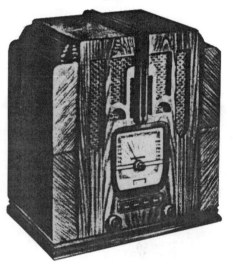

6052 CA1938
WOOD
$110

6070A CA1938
2-TONE WOOD
$45

6070 CA1938
WOOD
$45

6071 CA1938
2-TONE WOOD
$50

6072 CA1938
WOOD
$50

6073 CA1938
WOOD
$115

6130 CA1939
WOOD
$40

6201 CA1946
BAKELITE
$30

6231 CA1946
WOOD
$20

6261 CA1939
WOOD
$40

6262 CA1939
WOOD
$45

6321 CA1939
WOOD
$40

6235 CA1939
WOOD
$50

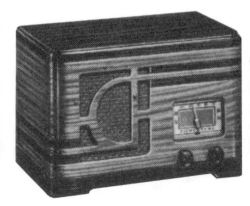

6350 CA1940
2-TONE WOOD
$35

6354 CA1939
2-TONE WOOD
$40

6356 CA1940
WOOD
$40

6359 CA1940
WOOD
$50

6362 CA1940
WOOD
$75

6421 CA1940
WOOD
$40

6425 CA1940
WOOD
$55

7001 CA1957
PLASTIC
$35

7003 CA1957
PLASTIC
$15

7006 CA1957
PLASTIC
$20

7010 CA1956
PLASTIC
$15

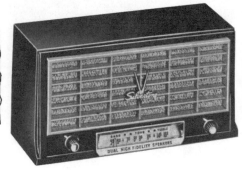

7013 CA1957
PLASTIC
$20

7020 CA1957
PLASTIC CLOCK-RADIO
$65

7021 CA1947
BAKELITE
$70

7025 CA1957
PLASTIC CLOCK-RADIO
$25

7034 CA1942
WOOD
$30

7037 CA1942
WOOD
$30

7039 CA1942
WOOD
$60

7054 CA1947
WOOD & TENITE
$40

7101 CA1941
WOOD
$25

7105 CA1942
WOOD
$25

7107 CA1942
WOOD
$25

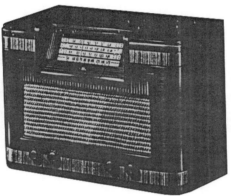

7111 CA1947
WOOD
$25

7204 CA1957
PLASTIC
$15

7221 CA1947
WOOD
$15

8000 CA1948
BAKELITE
$70

8003 CA1949
METAL (BY ARVIN)
$45

8005 CA1947
BAKELITE
$30

8005 CA1957
PLASTIC CLOCK-RADIO
$15

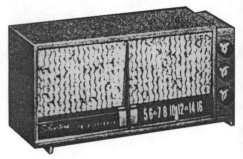

8009 CA1957
PLASTIC
$10

8017 CA1958
PLASTIC CLOCK-RADIO
$30

8019 CA1957
PLASTIC CLOCK-RADIO
$20

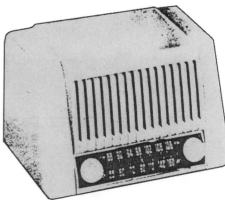

8022 CA1959
PLASTIC
$20

8050 CA1947
WOOD
$30

8052 CA1947
WOOD
$20

8221 CA1949
WOOD
$10

8231 CA1949
WOOD
$15

9002 CA1958
PLASTIC
$50

9004 CA1958
PLASTIC
$10

9007 CA1958
PLASTIC
$25

9011 CA1958
PLASTIC
$10

9018 CA1958
PLASTIC CLOCK-RADIO
$15

9021 CA1958
PLASTIC CLOCK-RADIO
$15

9027 CA1958
PLASTIC CLOCK-RADIO
$10

9260 CA1948
PLASTIC
$80

SIMPLEX

42 CA1935
2-TONE WOOD
$125

D CA1937
WOOD
$160

G CA1937
WOOD
$190

N CA1933
WOOD
$275

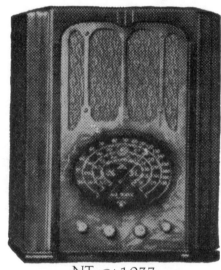

NT CA1937
WOOD
$150

P DELUXE CA1933
WOOD WITH MARQUETRY INLAY
$225

P CA1933
WOOD
$225

P MODERN CA1933
2-TONE WOOD
$350

Q CA1933
WOOD
$300

R Deluxe CA1933
Wood
$300

R CA1933
Wood
$275

RJ CA1937
Wood
$125

U Deluxe CA1933
Wood
$250

U CA1933
Wood
$175

V Deluxe CA1933
Wood
$275

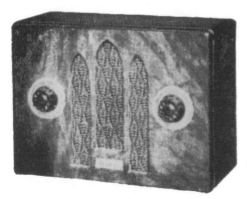

V CA1933
Wood
$90

W CA1933
Wood
$275

W 'World' CA1935
Wood
$235

SIMPLEX

WORLD CA1933
WOOD
$275

X CA1935
2-TONE WOOD
$175

Z CA1937
WOOD
$85

CA1930
WOOD
$250

CA1938
WOOD
$150

12 CA1939
WOOD, INGRAHAM CABINET
$125

44 CA1940
WOOD, INGRAHAM CABINET
$90

45 CA1940
WOOD WITH TENITE FACE, INGRAHAM CABINET
$110

46 CA1940
WOOD
$55

48 CA1940
PAINTED BAKELITE
$150

52 CA1940
WOOD
$40

53 CA1940
WOOD, INGRAHAM CABINET
$50

54 CA1940
WOOD, INGRAHAM CABINET
$65

62 CA1940
PAINTED BAKELITE
$70

63 CA1940
WOOD, INGRAHAM CABINET
$110

100 'TEENY WEENY' CA1939
IVORY PLASKON
$165

102 CA1939
BAKELITE
$150

102 CA1950
2-TONE PLASTIC
$45

105 CA1940
PAINTED BAKELITE
$75

108 CA1940
PAINTED BAKELITE
$90

172 'CLIPPER' CA1942
PAINTED BAKELITE
$50

176 CA1941
PAINTED BAKELITE
$45

178 CA1942
2-TONE WOOD
$55

179 CA1942
WOOD, INGRAHAM CABINET
$50

180 CA1942
WOOD, INGRAHAM CABINET
$95

181 CA1942
WOOD
$40

183 CA1942
WOOD
$40

210 CA1941
WOOD
$35

222 CA1948
PAINTED BAKELITE
$55

223 CA1941
WOOD
$40

240 CA1950
PAINTED BAKELITE
$120

246 'NIGHTENGALE' CA1941
PAINTED BAKELITE WITH READING LIGHT (BED)
$90

299 CA1950

Plastic

$40

306 CA1950

Wood

$20

618 CA1957

Colored Plastic

$45

CA1934

Wood

$70

CA1939

Wood with Black Lacquer Trim

$50

5 CA1931
WOOD
$225

5 CA1932
WOOD
$225

10 CA1932
WOOD
$160

57 CA1934
WOOD
$110

58 CA1934
WOOD
$100

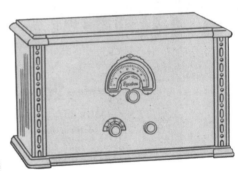

60 SW CONVERTER CA1932
WOOD
$120

65 CA1935
WOOD
$100

67 CA1935
WOOD
$130

70 CA1935
WOOD
$130

Sparton

100 CA1948
PAINTED BAKELITE
$35

132 'FOOTBALL' CA1950
PAINTED BAKELITE
$120

409 'SEVEN-SIDED' CA1938
MIRROR
BLUE $2000+   PEACH $2500+

410 'JUNIOR' CA1931
WOOD WITH REPWOOD GRILLE
$150

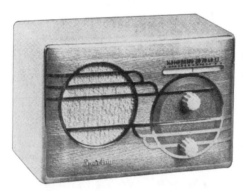

500 'CLOISONNE' CA1939
CATALIN CABINET w/COPPER ENAMEL FACE
2500+

506 'BLUEBIRD' CA1936
MIRROR & CHROME
BLUE $2300   PEACH $3000

517 CA1937
WOOD
$225

517W CA1937
IVORY LACQUER WOOD
$250

518 CA1938
WOOD
$250

527-2 CA1937
BLACK LACQUER WOOD WITH CHROME TRIM
$500+

528-2 CA1938
WOOD
$125

538 CA1938
WOOD
$125

557 'SLED' CA1938
MIRROR, WOOD $ CHROME (TEAGUE DES.)
BLUE $1700+   PEACH $2200+

558 '4-KNOB SLED' CA1938
MIRROR, WOOD $ CHROME (TEAGUE DES.)
BLUE $2000+   PEACH $2500+

608 CA1938
WOOD
$110

608B 'POLO CLUB' CA1938
BAKELITE: BROWN $350; BLACK $400
PLASKON: IVORY $500; RED $1500+

617 CA1937
WOOD
$150

628 CA1938
WOOD
$225

Sparton

638-6 CA1938
Wood
$210

716X CA1936
Wood with Marquetry & Black Lacquer
$350+

727 CA1937
Wood
$175

738 'Selectone' CA1938
Wood Clock-Timer-Radio
$325

5018 CA1938
Faux Grain Metal (Teague Des.)
$110

5518 CA1938
Faux Grain Metal (Teague Des.)
$80

6218 CA1939
Wood, Ingraham Cabinet
$500+

7140 (Canada) CA1939
Wood
$500+

'Easy-Goer' CA1954
Brightly Colored Plastic
$200

CA1934
WOOD WITH INSERT GRILLE
$350

CA1935
2-TONE WOOD
$175

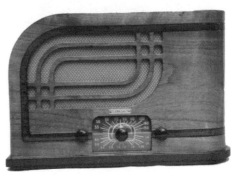

CA1936
2-TONE WOOD
$225

K CA1937
WOOD
$150

1-521 CA1940
WOOD
$85

1-531 CA1940
WOOD
$80

1-611 CA1940
WOOD
$80

3-5C1 CA1940
WOOD
$50

3-5E1 CA1940
BAKELITE
$65

3-5R7 CA1940
WOOD, INGRAHAM CABINET
$175

5R3 'CAMPUS' CA1941
IVORY PLASKON
$165

5R4 CA1941
WOOD, IM\NGRAHAM CABINET
$125

5R5 CA1941
WOOD, INGRAHAM CABINET
$175

5R6 CA1941
WOO, INGRAHAM CABINET
$110

5R7 CA1941
WOOD, INGRAHAM CABINET
$175

5S1 CA1941
WOOD, INGRAHAM CABINET
$135

5S2 CA1941
WOOD, INGRAHAMCABINET
$125

5U2 CA1941
WOOD, INGRAHAMCABINET
$115

5U3 CA1941
WOOD, INGRAHAMCABINET
$125

5W1 CA1941
WOOD
$45

6G1 CA1941
WOOD
$50

6J1 CA1941
BAKELITE
$65

6K1 CA1941
WOOD
$45

6N1 CA1941
WOOD
$55

6P1 CA1941
WOOD, INGRAHAM CABINET
$90

6P3 CA1941
WOOD, INGRAHAMCABINET
$125

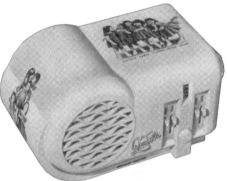

7-5B3Q 'DIONNE QUINTS' CA1940
IVORY PAINTED BAKELITE W/DECALS
$600

7-5R5 CA1940
WOOD, INGRAHAMCABINET
$175

7-513Q 'DIONNE QUINTS' CA1940
IVORY PAINTED BAKELITE W/DECALS
$500

7-514 CA1940
WOOD, INGRAHAMCABINET
$275

91-51 'SPADE' CA1940
WOOD, INGRAHAMCABINET
$500+

91-511 CA1939
WOOD
$45

91-512 CA1939
WOOD, INGRAHAMCABINET
$275

91-531 CA1939
WOOD WITH BLACK LACQUER TRIM
$135

97-52A CA1940
WOOD
$45

206 CA1946
WOOD
$40

1107 CA1934
WOOD
$250

1118 'LIBRARY' CA1934
CANVASBOARD
$250

1161 CA1934
WOOD
$150

1163 CA1934
WOOD
$165

1164 CA1934
WOOD
$200

1191 'RODNEY' CA1933
WOOD
$225

1236 'CAVALIER' CA1933
WOOD
$225

1301 CA1935
WOOD
$275

1451 CA1937
WOOD
$120

1461 CA1937
WOOD
$130

1671 CA1937
WOOD
$65

1691 CA1937
WOOD
$115

1711 CA1937
WOOD
$60

1731 CA1937
WOOD
$130

1811 CA1938
WOOD
$110

1821 CA1938
WOOD, INGRAHAM CABINET
$225

1911 CA1938
WOOD
$60

1921 CA1937
WOOD, INGRAHAM CABINET
$225

3043 '3-POSITION' CA1938
WOOD
$85

9001C CA1942
WOOD
$45

9002B CA1942
BAKELITE
$30

9003B CA1942
WOOD
$30

9005A CA1942
WOOD
$35

9014E CA1942
CATALIN
$600+

9151A CA1948
BAKELITE/PLASKON
$30

9152A CA1948
BAKELITE
$25

9153A 'Turnabout' CA1949
BAKELITE
$30

9162 CA1952
2-TONE PAINTED BAKELITE CLOCK-RADIO
$85

9182 CA1954
2-TONE PAINTED PLASTIC
$85

A-6-1Q 'DIONNE QUINTS' CA1940
IVORY PAINTED BAKELITE WITH DECALS
$500+

A-72-T3 CA1948
WOOD WITH METAL GRILLE
$25

'APARTMENT' CA1932
WOOD
$500

'GOOD COMPANION' CA1937
BLACK LACQUER WOOD & METAL
$1200+

SW CONVERTER CA1931
WOOD
$150

'METROPOLITAN' CA1931
WOOD
$375

R-110A CA1932
WOOD (10 TUBES)
$325

R-1361A CA1937
WOOD
$250

'TABLE' CA1932
WOOD
$225

CA1933
METAL
$350

'FERRODYNE' CA1937
WOOD
$250

58T CA1936
2-TONE WOOD
$140

69 'SW CONVERTER' CA1935
2-TONE WOOD WITH BLACK LCQUER TRIM
$250

126H CA1938
WOOD
$145

235H CA1937
WOOD
$225

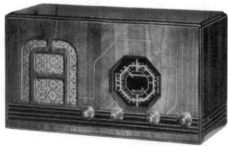

322C CA1937
WOOD
$140

322H CA1937
WOOD
$165

323H CA1937
WOOD
$115

325J CA1938
WOOD
$130

335H CA1939
WOOD
$145

337H CA1937
WOOD
$120

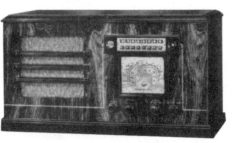

340H CA1937
WOOD
$120

354H CA1938
WOOD
$140

400H CA1940
WOOD
$90

410H CA1941
WOOD
$80

412H CA1940
WOOD WITH BLACK LACQUER TRIM, INGRAHAM CAB.
$145

420J CA1940
WOOD
$85

500H CA1941
BAKELITE
$55

520H CA1941
WOOD
$90

951H CA1938
WOOD
$90

952H CA1938
WOOD
$80

972H CA1938
WOOD
$85

992H CA1938
WOOD
$110

1051 (CANADA) CA1939
PAINTED BAKELITE
$225

1101H CA1947
2-TONE WOOD
$45

1121HW CA1947
WOOD
$50

1204 CA1950
WOOD
$45

1210 CA1950
WOOD
$30

1210HW CA1948
WOOD
$30

'TE-LEK-TOR-ET' REMOTE CA1933
WOOD
$150

'TE-LEK-TOR-ET' SPEAKER CA1933
WOOD WITH BLACK LACQUER TRIM
$350+

109 CA1947
BAKELITE
$50

109W CA1947
BAKELITE
$40

131 CA1947
WOOD
$30

135 CA1947
BAKELITE
$45

136 CA1947
WOOD, INGRAHAM CABINET
$60

139 CA1947
WOOD, INGRAHAM CABINET
$250

156 CA1948
MARBELED PLASTIC
$75

157 CA1948
PLASTIC
$50

158 CA1949
BAKELITE
$45

165 CA1949
MARBELIZED COLORED PLASTIC
$125

184 CA1949
PLASTIC
$35

190 CA1949
PLASTIC
$50

198 CA1949
WOOD
$25

228 CA1951
COLORED PLASTIC
$65

230 CA1951
PLASTIC
$45

TR70 CA1950
COLORED PLASTIC
$65

53 CA1935
WOOD
$140

60A CA1935
WOOD WITH CHROME TRIM
$175

63 CA1935
WOOD
$130

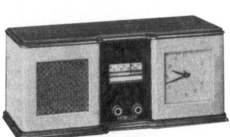

325 CA1940
BLONDE WOOD WITH BLACK LACQUER TRIM
$75

431 CA1937
WOOD
$60

502 CA1937
WOOD
$110

512 CA1936
WOOD
$350

522 CA1937
WOOD
$110

525 CA1937
WOOD
$225

527 CA1938
Wood
$60

549 CA1937
Wood
$85

627 CA1938
Wood
$70

635 CA1937
Wood
$70

701 CA1937
Wood with Black Lacquer Trim
$170

702 CA1937
Wood
$165

703 CA1937
Wood
$140

730 CA1937
Wood
$60

5055 CA1948
Bakelite with Tenite Grille
$60

2401 CA1947
2-Tone Wood
$30

2405 CA1947
2-Tone Wood
$30

2501 CA1947
Bakelite
$45

2531 CA1947
Painted Bakelite
$70

2533 'Bullet' CA1947
Bakelite
$135

2565 CA1947
Wood
$25

2612 CA1947
Painted Bakelite
$25

2613 CA1947
Wood
$30

2701 CA1947
Wood
$40

5 CA1935
PEACH, BLUE, GREEN OR SILVER MIRROR
$1500+

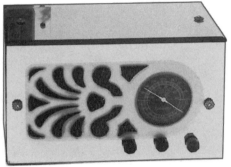

5X CA1935
PEACH, BLUE, GREEN OR SILVER MIRROR
$1500+

44 CA1934
2-TONE WOOD WITH BLACK LACQUER TRIM
$225

45M CA1938
PEACH, BLUE, GREEN OR SILVER MIRROR
$1000+

75 CA1936
WOOD
$175

100 CA1938
WOOD
$160

113AW CA1939
WOOD WITH BLACK LACQUER TRIM
$90

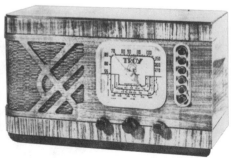

825 CA1939
WOOD
$120

'HI-MU PENTODE MIDGET' CA1932
WOOD
$275

TRUETONE

D702 CA1937
BLACK BAKELITE WITH CHROME GRILL BARS
$225

D713 'AMBASSADOR' CA1939
WOOD
$55

D715 'CHALLENGER' CA1939
WOOD
$65

D716 CA1939
WOOD
$70

D724 CA1937
WOOD
$75

D730 'PLA-MOR' CA1939
BAKELITE
$75

D909 'PETITE' CA1941
BROWN BAKELITE $120   IVORY PLASKON $225
GREEN OR RED PLASKON $600+

D911 'SUPREME' CA1940
WOOD WITH BLACK LACQUER TRIM
$55

D914 CA1941
WOOD
$60

D915 'CORONET' CA1940
BEETLE PLASTIC
$350

D934 'DELUXE' CA1940
WOOD
$50

D935 'MASTER' CA1940
WOOD
$40

D1001 'SENTINEL' CA1941
WOOD, INGRAHAM CABINET
$160

D1002 'GLOBETROTTER' CA1941
WOOD
$70

D1003 'ENVOY' CA1941
WOOD
$65

D1011 CA1941
PAINTED BAKELITE
$75

D1015 'CORONET' CA1941
PAINTED BAKELITE
$135

D1020 CA1941
WOOD
$30

D1021 CA1941
WOOD
$40

D1022 CA1941
WOOD
$50

D1035 CA1941
WOOD
$40

D1224 CA1940
WOOD, INGRAHAM CABINET
$125

D2613 CA1947
PAINTED BAKELITE
$40

D2615 'VICTORY' CA1942
CANVASBOARD
$50

D2620 CA1947
WOOD
$60

D2621 CA1947
WOOD
$55

D2624 CA1947
WOOD
$25

D2644 CA1946
WOOD
$25

D2663 CA1946
WOOD
$25

D2665 CA1948
FAUX GRAIN METAL WITH CHROME TRIM
$50

D2693 CA1948
PAINTED BAKELITE
$50

D2718 CA1948
PAINTED BAKELITE
$70

D2762 CA1948
BAKELITE
$25

D2815 CA1948
BAKELITE
$65

D2810 CA1948
BAKELITE
$20

'SUPREME' CA1939
WOOD
$110

7A C1932
WOOD
$225

7D CA1932
REPWOOD
$300

8A CA1932
WOOD
$325

9A CA1932
REPWOOD
$300

20 'GLORIETTE' CA1932
WOOD
$280

24 CA1933
WOOD
$225

25A CA1932
WOOD
$50

26P 'GLORITONE' CA1931
WOOD
$250

27A 'GLORITONE' CA1931
WOOD
$210

27 'GLORITONE' CA1931
WOOD
$175

27 'MARFIELD' CA1931
WOOD
$225

32A CA1932
WOOD
$275

3072 CA1932
WOOD
$225

3082 CA1933
WOOD WITH MARQUETRY INLAY
$90

3092 CA1933
WOOD WITH MARQUETRY INLAY
$90

CA1931
WOOD
$190

6PT-TX (Canada) ca1956
Bakelite with Gold Plastic Grille (Plug-In Transistors)
$225

22 ca1934
Wood
$115

27 ca1934
Wood
$80

327-T6 ca1951
Bakelite with Plastic Grille
$25

345-T5 ca1951
Bakelite
$25

350-T7 ca1951
Wood
$25

356-T5 ca1951
Painted Bakelite Clock-Radio
$50

'Columnette' ca1932
Wood
$110

H-342-P5U ca1951
Black & Red Urea
$325

H-404-T5 CA1953
COLORED PLASTIC CLOCK-RADIO
$40

H-583-T5 CA1958
COLORED PLASTIC CLOCK-RADIO
$85

H-679-T4 CA1959
COLORED PLASTIC CLOCK-RADIO
$45

H-686-P8 CA1959
COLORED PLASTIC MINI-CLOCK-RADIO (TRANSISTOR)
$125

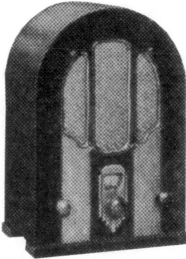

'PORTA-MIDGET' CA1932
2-TONE WOOD
$175

WR-12-X12 CA1942
BAKELITE
$30

WR-28 CA1934
WOOD
$150

WR-101 CA1936
WOOD
$145

WR-102 CA1936
WOOD
$40

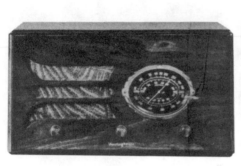

WR-103 CA1937
WOOD
$45

WR-165 CA1938
WOOD
$275

WR-166 CA1938
IVORY PLASKON WITH RED PLASKON TRIM
$225

WR-168-B CA1939
WOOD
$25

WR-169 CA1939
WOOD
$50

WR-173 CA1939
BAKELITE
$90

WR-175 CA1939
IVORY PAINTED BAKELITE
$45

WR-201 CA1936
WOOD
$120

WR-203 CA1936
WOOD
$135

WR-204 CA1936
Wood with Black Lacquer Trim
$160

WR-205 CA1936
Wood with Marquetry Inlay
$175

WR-208 CA1936
Wood
$65

WR-228 CA1938
Wood
$70

WR-258 CA1939
2-Tone Wood
$40

WR-272-L CA1939
Wood
$35

WR-290 CA1941
Wood
$30

WR-470 CA1942
Wood
$35

WR-603 CA1937
Wood
$80

2VB7-67 'Carillion' ca1934
Wood
$275

3J5-55 'Room-Mate' ca1934
Wood with Marquetry Inlay
$110

3S5-66 'Cameo II' ca1934
Wood
$140

A Series ca1938
Wood
$150

A-33 ca1938
Wood
$175

A-51 ca1939
Ivory Plaskon
$350

Wall Radio ca1938
Ivory painted Wood with Natural Trim
$125

Wall Radio ca1938
Painted Metal
$225

ca1937
2-Tone Wood
$450

CA1939
Wood
$40

CA1939
Wood with Black Lacquer Trim
$125

CA1939
Wood
$45

CA1939
Wood
$150

489 CA1937
WOOD
$65

833 CA1938
WOOD
$60

'CHALLENGER' CA1939
WOOD
$50

'JUNIOR' (MISSION BELL) CA1939
BAKELITE
$250

'SUPERHET SEVEN' CA1939
WOOD
$75

5 CA1957
COLORE PLASTIC
$40

11 CA1946
BAKELITE
$35

14 CA1946
BAKELITE
$65

15 CA1946
BAKELITE WITH BRASS TRIM
$45

23 CA1946
PIANTED BAKELITE
$45

27 CA1946
BLONE WOOD WITH BLACK LACQUER TRIM
$45

28 CA1936
WOOD
$450

29 CA1946
WOOD
$45

30 CA1946
WOOD
$60

34 CA1946
WOOD
$50

35 CA1946
WOOD
$40

106 CA1937
WOOD (REC. VEHICLE-SPEAKER SEPARATE)
$150 ($250 WITH SPEAKER)

107 CA1937
WOOD (REC. VEHICLE-SPEAKER SEPARATE)
$225 ($400 WITH SPEAKER)

118 CA1937
WOOD
$135

132 CA1937
WOOD
$160

210 CA1933
WOOD
$350

217 CA1938
WOOD
$175

219 CA1938
WOOD
$175

220 CA1938
WOOD
$275

221 CA1938
WOOD
$300

222 CA1938
WOOD
$175

223 CA1938
WOOD
$150

227 CA1938
WOOD
$80

230 CA1933
WOOD
$400

230 CA1938
WOOD WITH FULL FAUX FINISH
$140

231 CA1938
WOOD
$80

233 CA1938
WOOD
$175

250 CA1933
WOOD
$325

288 CA1934
WOOD
$350

313 'BULLET' CA1939
BAKELITE
$145

316 CA1939
WOOD
$65

322 CA1939
WOOD
$160

323 CA1939
WOOD
$190

330 CA1939
WOOD WITH FULL FAUX FINISH
$155

416 CA1940
BAKELITE
$125

417 CA1940
IVORY PAINTED BAKELITE
$125

418 CA1940
BAKELITE
$140

422 'BULLET' CA1940
BAKELITE
$145

426 CA1940
WOOD
$160

427 CA1940
WOOD
$225

429 CA1940
WOOD
$65

433 'ZEPHYR' CA1940
WOOD
$250

434 CA1940
WOOD
$350

435 CA1940
2-TONE WOOD
$50

436 CA1940
2-TONE WOOD
$165

438 CA1940
WOOD
$175

441 CA1940
WOOD
$135

442 CA1940
WOOD
$140

443 CA1940
WOOD
$165

510 CA1941
BAKELITE
$50

511 CA1941
BAKELITE
$65

515 CA1941
BAKELITE
$110

520 CA1941
PAINTED BAKELITE
$55

527 CA1941
WOOD
$90

529 CA1941
WOOD
$160

530 CA1941
WOOD
$175

532 CA1941
WOOD
$160

533 CA1941
WOOD
$40

535 CA1941
WOOD
$45

536 CA1941
WOOD
$45

537 CA1941
WOOD
$160

610 CA1942
PAINTED BAKELITE
$50

611 CA1942
BAKELITE
$60

612 CA1942
PAINTED BAKELITE
$65

614 CA1942
PAINTED BAKELITE
$70

615 CA1942
WOOD
$65

620 CA1942
WOOD
$90

625 CA1942
WOOD
$110

627 CA1942
WOOD
$120

627 CA1942
WOOD
$60

629 CA1942
WOOD
$90

630 CA1942
WOOD
$65

631 CA1942
BAKELITE
$65

632 CA1942
WOOD
$90

633 CA1942
WOOD
$95

634 CA1942
WOOD
$90

705 CA1934
WOOD
$100

706 CA1934
WOOD
$120

712 CA1934
WOOD
$450

715 CA1934
WOOD
$550

801 CA1935
WOOD
$125

805 CA1935
WOOD
$325

825 CA1935
WOOD
$135

827 CA1935
WOOD WITH CHROME TRIM
$750+

908 CA1935
WOOD
$400

918 CA1950
BAKELITE
$35

920 CA1950
BAKELITE
$25

921 CA1950
BAKELITE
$20

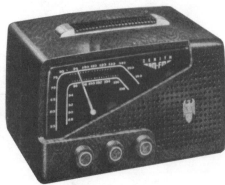

922 CA1950
BAKELITE
$20

AH CA1931
WOOD
$350

G510 CA1950

BAKELITE WITH METAL GRILLE

$40

G511 CA1950

BAKELITE WITH METAL GRILLE

$40

G516 CA1950

IVORY PLASKON

$75

G615 CA1950

BAKELITE WITH METAL GRILLE

$40

G725 CA1950

BAKELITE

$20

H511 CA1951

BAKELITE

$45

H715 CA1950

BAKELITE WITH METAL GRILLE

$35

H723 CA1950

BAKELITE

$15

LP CA1931

WOOD

$375

H5O9V CA1956
COLORED PLASTIC
$45

Y519 CA1956
COLORED PLASTIC CLOCK-RADIO
$50

Y519A CA1956
COLORED PLASTIC
$45

'BROADWAY' CA1955
COLORED PLASTIC
$45

'CARNIVAL' CA1955
COLORED PLASTIC
$25

'CLIPPER' CA1955
COLORED PLASTIC
$35

EMPRESS CA1959
COLORED PLASTIC CLOCK-RADIO
$15

'RADIO NURSE' (NOGUCHI DES.) CA1942
BAKELITE
$900+

'SUPER INTERLUDE' CA1959
WOOD WITH PLASTIC FACE
$10

'SUPER SYMPHONAIRE' CA1955
WOOD
$15

'TIP-TOP HILIDAY' CA1950
PLASTIC
$50

'UNIVERSAL' CA1950
LEATHERETTE
$45

'VICEROY' CA1949
WOOD
$25

'ZENETTE' CA1931
WOOD
$400

'ZENETTE' CA1950
COLORED PLASTIC
$65

'ZEPHYR' CA1959
PLASTIC
$20

'ZETOVOX'(EXPORT) CA1932
WOOD
$500+

ABC
CA1959
PINK PLASTIC
$30

ADVANCE
J CA1931
WOOD
$325

ADVANCE
MIDGET CA1931
WOOD
$275

AERO
MIDGET CA1932
WOOD
$225

AERO
SUPERPHONIC 6 CA1933
WOOD
$150

AERO
SUPERPHONIC 6 CA1933
WOOD
$250

AIR CHIEF
PATRIOT CA1942
RED, WHITE & BLUE PAINTED BAKELITE
$125

AIR CHIEF
CA1935
WOOD
$130

AIR CASTLE
'TULIPS' CA1937
2-TONE WOOD
$400+

AIR CASTLE
'WILTING TULIPS' CA1937
WOOD
$135

AIR MASTER
CA1937
PAINTED METAL
$90

ALTMAN
609 CA1937
WOOD
$160

AMERICAN
MIDGET CA1932
WOOD
$240

AMERICAN
6K WALL RADIO CA1947
CATALIN
$120

APPROVED
FM CONVERTER CA1948
WOOD
$40

ARIA
6A265 CA1940
WOOD
$65

ARISTA
L82 CA1947
WOOD
$35

ARKAY
CA1934
WOOD WITH CHROME RING IN GRILLE
$400+

ARKAY
CA1934
BLONDE WOOD WITH BLACK LACQUER & CHROME GRILLE
$350+

ARLINGTON
501 CA1935
2-TONE WOOD
$150

ATCHISON
MIDGET CA1932
WOOD
$225

ATOMIC
ELECTRO (KIT) CA1947
WOOD
$35

AUTOCRAT
101 CA1939
IVORY PLASKON
$250

AUTOCRAT
24BG CA1933
TONED WOOD
$150

AUTOCRAT
MIDGET CA1932
TONED WOOD
$190

BALDWIN
50/51 'BALDWINETTE' CA1930
WOOD
$225

BALDWIN
70/71 'CONSOLETTE' CA1930
WOOD
$225

BALKEIT
42E 'JR' CA1933
TONED WOOD
$175

BALKEIT
44 CA1933
PAINTED CANVASBOARD
$200

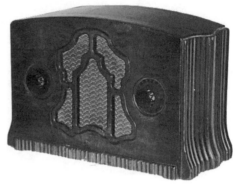

BALKEIT
65S CA1933
WOOD
$170

BEVERLY
CA1933
2-TONE WOOD
$250

BOWERS
T-61 CA1947
WOOD
$25

BULOVA
120 CA1957
COLORED PLASTIC CLOCK-RADIO
$60

BULOVA
204 CA1955
COLORED PLASTIC W/REVERSE-PAINTED TRIM
$80

BULOVA
CA1959
COLORED PLASTIC W/REVERSE PAINTED GRILLE
$60

CAMDEN
81 CA1935
WOOD
$115

CAMDEN
91 CA1935
WOOD WITH MARQUETRY INLAY
$125

CARDINAL
'CINDERELLA' CA1932
WOOD WITH BLACK LACQUER INSERT GRILLE
$325

CASE
510 CA1936
WOOD
$225

CASE
700 CA1936
WOOD
$140

CAVALCADE
RS1A CA1947
MARBELED PLASTIC
$150

CAVALIER
LK-447 CA1933
2-TONE WOOD
$170

CAVALIER
SF-547 CA1933
2-TONE WOOD
$180

CBS
5158 CA1958
PLASTIC
$45

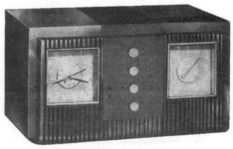

CBS
CLOCK-RADIO CA1951
BAKELITE
$35

CHAINWAY
781 CA1936
WOOD
$75

CHAMPION
CA1934
METAL
$70

CHANNEL MASTER
6532 CA1958
COLORED PLASTIC
$30

CHURCHILL
'GUTMANN' CA1940
LEATHER COVERED WOOD
$125

CLINTON
CA1934
WOOD WITH CHROME GRILLE INSERT
$350+

CLINTON
CA1937
WOOD WITH BLACK LACQUER TRIM
$225

CLUB
'COUNTRY CLUB' CA1933
PAINTED WOOD WITH METAL INSERT GRILLE
$400+

COLUMBIA
C-31 CA1932
WOOD
$210

COLUMBIA
C-81 CA1932
2-TONE WOOD
$325

COMMONWEALTH
170 CA1933
WOOD
$220

CONTINENTAL
'PIANO' CA1940
DARK GREEN BAKELITE
$375

COOP
T-65 CA1937
WOOD
$55

CORONA
CA1936
WOOD
$120

CROYDON
136 CA1937
WOOD
$110

CROYDON
CA1932
WOOD
$275

CRUSADER
CA1936
WOOD
$35

CRUSADER
CA1941
BAKELITE
$50

CUNNINGHAM
CA1936
2-TONE WOOD
$110

CYARTS
B CA1947
LUCITE/PLEXIGLASS
$1000+

DAVISON-HAYNES
MIDGET CA1931
WOOD
$240

DEFOREST-CROSLEY (CAN)
CA1932
WOOD WITH MARQUETRY INLAY
$350

DIAMOND
CA1936
WOOD
$110

DEFOREST RADIO INSTITUTE
CA1937
FLOCKED WOOD WITH CRYSTAL FINISH METAL TRIM
$750+

DUMONT
RA346 CA1956
WOOD WITH PLASTER FACE
$115

ECA
102 CA1947
BAKELITE
$35

ECA
107 CA1947
WOOD
$40

ECA
132 CA1947
WOOD
$30

337

ECA
2O1 CA1947
2-TONE WOOD
$55

EH
CA1935
WOOD
$160

EMERALD
5O1 CA1947
MIRROR OVER WOOD
$750+

EMPIRE
30 CA1933
WOOD
$200

EMPIRE
AA 'SPORTSMAN' CA1937
WOOD
$175

EMPIRE
CA1933
WOOD
$145

ENVOY
'LONGFELLOW' CA1932
WOOD
$210

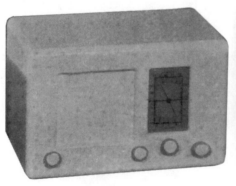

EPSEY
771AW CA1938
IVORY LACQUER WOOD
$55

EPSEY
'CHINESE LACQUER' CA1938
HAND-PAINTED WOOD
$225

EPSEY
E CA1935
WOOD
$140

EPSEY
CA1938
IVORY LEATHER OVER WOOD
$225

ERLA
22P CA1931
WOOD CLOCK-RADIO
$325

ERLA
72 CA1930
WOOD CLOCK-RADIO
$300

ERLA
SW CONVERTER CA1933
WOOD
$165

FALCK
88 CA1931
WOOD
$225

FALCK
E CA1931
WOOD
$250

FARADAY
C259 'REFRIGERATOR' CA1947
WOOD
$250

FERGUSON
CA1935
WOOD
$90

FERGUSON
238 CA1938
WOOD
$90

FERGUSON
BL60 CA1932
WOOD
$350

FLINT
MIDGET CA1931
WOOD
$250

FLINT
MIDGET CA1931
WOOD WITH REPWOOD GRILLE
$250

FORDSON
6T 'GOLDENTONE' CA1933
WOOD
$225

FRANKLIN
55CU CA1934
WOOD
$150

FUTURAMIST
CA1958
COLORED PLASTIC
$45

GENERAL
CA1933
WOOD
$130

GENERAL
CA1934
WOOD
$150

GENERAL
J CA1937
WOOD
$200

GLOBE
95 CA1948
WOOD WITH METAL HORSE
$250

GRANCO
730 CA1956
PLASTIC
$35

GRANTLINE
508A CA1947
PAINTED BAKELITE
$175

GRIFFIN-SMITH
MIDGET CA1931
WOOD
$200

GUILD
380T 'TOWN CRIER' CA1958
WOOD
$110

GUILD
484 'SPICE RACK' CA1956
WOOD
$65

GUILD
'OLD TYMER' CA1958
WOOD
$90

HALLICRAFTERS
EC102 CA1946
BAKELITE
$60

HALLICRAFTERS
EC-103 CA1946
WOOD
$45

HALLICRAFTERS
612 CA1954
COLORED PLASTIC
$110

HALLICRAFTERS
AT-1 CA1952
PLASTIC
$60

HARFIELD
CA1933
WOOD
$140

HIBBARD
KM-450 CA1940
WOOD, INGRAHAM CABINET
$125

HORN
CA1936
WOOD WITH MIRROR TILES
$750+

HUDSON
CA1934
WOOD
$225

ICA
SW CONVERTER CA1933
WOOD
$90

IMPERIAL
25 CA1937
WOOD WITH BLACK LACQUER TRIM
$110

IMPERIAL
629 (10 TUBES) CA1938
WOOD
$500+

JAX
4 CA1933
WOOD
$175

JENKINS
JD-30 CA1932
WOOD
$325

JENNINGS
COIN-OP CA1932
WOOD
$250

JESSE FRENCH
MIDGET CA1931
WOOD
$300

KELLER-FULLER
RADIETTE 14 CA1931
WOOD
$200

KELLER-FULLER
RADIETTE CA1930
WOOD
$250

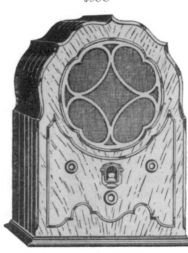

KENNEDY
CORONET CA1931
WOOD
$225

KENNEDY
CORONET CA1932
WOOD
$275

KENNEDY
IMPERIAL CA1931
WOOD
$250

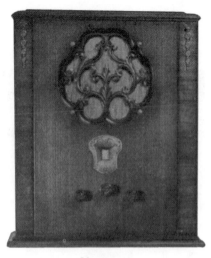

KOLSTER
K-60 CA1933
WOOD WITH REPWOOD GRILLE
$125

KOLSTER
K-110/114 CA1933
WOOD
$110

KRES-TONE
75-1 CA1948
WOOD
$45

LANG
CA1935
WOOD
$225

LARK
CA1933
TONED WOOD
$175

LEAR
561 CA1947
WOOD
$25

LEAR
562 CA1947
BAKELITE
$35

LEE
400 CA1948
WOOD WITH FLOCKING (FAUX SUEDE)
$110

LeWol
61 CA1934
WOOD WITH BLACK LACQUER TRIM
$350+

LeWol
S CA1933
WOOD WITH BLACK LACQUER TRIM
$175

LIFCO (CAN)
L-630 'ROAMER DELUXE' CA1953
BAKELITE
$125

LINCOLN
CA1938
WOOD
$40

LIND'S
CA1932
WOOD
$175

LITTLE GIANT
CA1934
2-TONE WOOD
$200

MAESTRO
5-TUBE AW CA1932
POT METAL
$500+

MAQUIRE
500D1 CA1948
PAINTED BAKELITE
$40

MANTOLA
8A CA1932
WOOD
$110

MARCONI (CAN)
42 CA1931
WOOD
$350+

MARCONI (CAN)
238 CA1948
WOOD
$40

MARCONI (CAN)
CA1935
WOOD WITH CHROME GRILLE
$450+

MASTER
70 'MIGHTY MIDGET' CA1930
WOOD
$175

MAYEROLA
CA1930
WOOD
$175

MAYFAIR
CA1932
WOOD
$225

MAY
533 CA1935
WOOD
$140

McCORMICK
'MANTLE MIDGET' CA1931
WOOD CLOCK-RADIO
$250

MELORAD
'CATHEDRALTONE' CA1931
WOOD
$250

MINERVA
UNIVERSAL CA1932
WOOD
$150

MIR-RAY
CA1935
MIRROR
$1200+

MIRROR-TONE
4B7 CA1947
BAKELITE
$70

MIRROR-TONE
804 CA1948
IVORY PLASKON
$70

MISSION BELL
387B 'TEMPO' CA1938
IVORY PLASKON WITH GOLD TRIM
$275

MISSION BELL
MIDGET CA1932
WOOD
$275

MISSION BELL
CA1936
WOOD
$180

MITCHELL
1254 'MADRIGAL' CA1951
BAKELITE
$50

MODERNAIRE
ONE-TUBE MIDGET
RED PLASKON
$300+

MOHAWK/LYRIC
SA65 CA1933
WOOD
$170

MONARCH
'PYRAMID' CA1934
2-TONE WOOD
$125

MUSIC MASTER
7-TUBE SUPERHET
WOOD
$450+

MUSIC MASTER
CLOCK-MIDGET CA1932
WOOD CLOCK-RADIO
$500+

MUSICAIRE
576 CA1948
WOOD
$65

MUSICAIRE
CA1955
COLORED PLASTIC
$50

MUSIQUE
4 CA1932
WOOD
$175

NAMCO
601 CA1946
CATALIN
$2000+

NATIONAL UNION
G619 CA1947
WOOD
$25

NATIONAL PFANSTIEHL
CA1931
WOOD
$180

NORTHERN ELECTRIC (CAN)
5508 'MIDGE' CA1948
BAKELITE
$135

OFFICIAL LEAGUE
TROPHY BASEBALL CA1948
BAKELITE
$750+

OLYMPIC
CA1958
PLASTIC
$45

OZARKA
93 CA1932
WOOD
$1200+

PEKO
MAESTRO CA1932
WOOD CLOCK-RADIO
$250

PENTATRON
CA1931
METAL CABINET WITH REPWOOD FACE
$300

PERRY
MULTI-MU CA1932
WOOD
$150

PHILHARMONIC
CLOCK-RADIO CA1951
COLORED PLASTIC
$60

PHONOLA (CAN)
40U51P3 CA1940
PAINTED WOOD
$35

PICKWICK
CA1934
WOOD
$85

PIN-UP CORP.
CA1957
SWIRLED PLASTIC
$175

PIONEER
SUPER-HETERODYNE CA1932
WOOD
$190

PLAZA
MIDGET CA1932
WOOD WITH REPWOOD GRILLE
$250

PLYMOUTH
MIDGET CA1931
WOOD
$190

PORTO-RADIO
'SMOKERETTE' CA1948
BAKELITE
$175

POST
SKYCHIEF CA1938
WOOD
$45

POWELL
MIDGET CA1931
WOOD
$225

PRATT
CA1936
MIRROR
$750

PRATT
CA1937
MIRROR
$400

PREMIER
712 CA1934
2-TONE WOOD
$80

PREMIERE
MIDGET CA1931
WOOD
$225

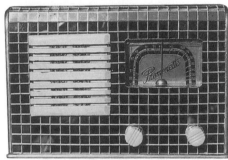

PROMONETTE
CA1946
WOOD WITH MIRROR TILE
$250

PURE
PURE OIL CO. PROMO RADIO
WOOD
$1500+

RADIO-CRON
A-1 CA1932
WOOD CLOCK-RADIO
$250

RADIO-CRON
K-10 CA1932
WOOD CLOCK-RADIO
$275

RADIO-CRON
K-20 CA1932
WOOD CLOCK-RADIO
$350

RADIO-GLO
CA1935
METAL, MIRROR & STAINED GLASS
$1200

RADIO LAMP OF AMERICA
CA1940
BRASS WITH GLASS BOWL (MISSING IN PHOTO)
$350

RCI (RADIO CHASSIS INSTITUTE)
CA1933
WOOD WITH MARQIETRY INLAY
$165

RCI (RADIO CHASSIS INSTITUTE)
6-TUBE SUPERHET CA1933
WOOD
$160

REMBERT
SW CONVERTER CA1932
WOOD
$90

REPORTER
CA1932
WOOD
$185

REPUBLIC
MIDGET CA1931
WOOD
$170

REVERE
399 CA1931
WOOD
$225

RK RADIO
RADIO KEG
WOOD/REPWOOD
$300

ROGERS (CAN)
R151 CA1950
PAINTED BAKELITE
$75

ROYAL
CA1935
WOOD
$150

ROYAL
CA1933
3-TONE WOOD
$120

ROYAL
AC/DC CA1933
WOOD
$135

RPC
AEOLIAN CA1946
IVORY PLASKON
$45

SATURN
CA1937
WOOD
$70

SEE-ALL
SCANNING DISK TV/AW RADIO CA1932
WOOD
$10,000+

SETCHELL-CARLSON
416 CA1946
IVORY & RED PLASKON
$80

SETCHELL-CARLSON
RADIO-INTERCOM CA1947
WOOD
$30

SHEFFIELD
CA1936
WOOD
$140

SILVER
159 CA1937
WOOD
$500+

SKY-HAWY
SL5D CA1934
WOOD WITH MARQUETRY INLAY
$125

SKYROVER
C CA1932
WOOD
$140

SPEAK-O-PHONE
RADIO-LITE CA1939
BAKELITE
$350

STANDARD
CA1949
COLORED UREA
$90

STEELMAN
5101 CA1951
WOOD
$20

STEELMAN
AF1100 'CORONET' CA1951
WOOD
$20

STEINITE
16 CA1932
WOOD
$250

STEIN
'AZTEC' CA1931
TONED WOOD
$300+

STERLING
F CA1931
WOOD
$165

STERLING
MIDGET CA1931
WOOD
$175

STUDEBAKER
'WREN' CA1932
WOOD
$300

SYLVANIA
563B CA1953
PLASTIC
$30

TCA
MIDGET CA1931
WOOD
$175

TELECHRON
8H67 CA1948
MARBELED UREA (PINK, BLUE, GREY)
$225

TEMPLETON
SUPER-TONE CA1932
WOOD
$190

TEMPLE
E514 CA1947
WOOD
$20

Temple
H411 ca1948
Metal
$90

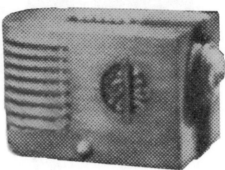

Tempo-Tone
ca1939
Wood
$500+

Tiffany-Tone
83W ca1937
Wood
$275

Tiny Tim
Pee Wee ca1939
Painted Bakelite
$150

Tom Thumb
ca1938
Black Bakelite with Chrome Trim
$750

Tom Thumb
ca1938
Wood
$350

Trutest
Midget ca1932
Wood
$220

Tun-O-Matic
ca1934
Wood Automatic Clock-radio
$500+

Ultradyne
Pee Wee ca1939
2-Tone Wood
$225

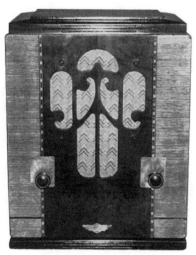

UNITED MOTORS
4054 CA1934
WOOD WITH MARQUETRY INLAY
$225

UNIVERSAL
7232 CA1936
WOOD
$100

UNIVERSAL
CA1932
2-TONE WOOD
$160

VITALTONE
MIDGET CA1930
TONED WOOD
$110

WAL-TONE
CA1932
2-TONE WOOD
$75

WARE
BANTAM CA1931
WOOD
$225

WARWICK
0-76 CA1948
BAKELITE
$135

WELLS GARDNER
1081A-704 CA1938
WOOD
$90

WELLS GARDNER
A11 CA1938
BAKELITE
$110

WESTERN
CA1937
WOOD
$130

WINGS CIGARETTES
PROMOTIONAL RADIO (BY RCA)
METAL CABINET OVER WOOD FRAME
$1700+

WOLPER & SEIDSCHER
'REPORTER' CA1931
WOOD
$190

WOODSTOCK
MIDGET CA1932
WOOD
$185

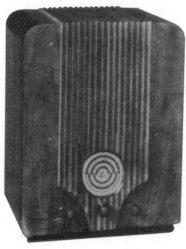

WORLD
583 CA1936
WOOD WITH BLACK LACQUER DETAIL
$250

'MUSIC BOX' CA1931
WOOD WITH INSERT GRILLE
$300

'CLARION' CA1931
WOOD WITH METAL GRILLE AND BEZEL AREA
$400+

'VITA-TONE' CA1931
WOOD WITH BLACK LACQUER INSERT GRILLE
$500+

NOTES

NOTES

NOTES

NOTES

NOTES